NANOTECHNOLOGY SCIENCE AND TECHNOLOGY

SAFE NANOTECHNOLOGY AND PROGRESS IN THE WORKPLACE (UPDATED)

NANOTECHNOLOGY SCIENCE AND TECHNOLOGY

Additional books in this series can be found on Nova's website under the Series tab.

Additional E-books in this series can be found on Nova's website under the E-books tab.

NANOTECHNOLOGY SCIENCE AND TECHNOLOGY

SAFE NANOTECHNOLOGY AND PROGRESS IN THE WORKPLACE (UPDATED)

DOMINICK A. CAGGIONE
EDITOR

Nova Science Publishers, Inc.
New York

For permission to use material from this book please contact us:
Telephone 631-231-7269; Fax 631-231-8175
Web Site: http://www.novapublishers.com

Library of Congress Cataloging-in-Publication Data

Safe nanotechnology and progress in the workplace (updated) / editor, Dominick A. Caggione.
p. cm.
Previously published in: DHHS (NIOSH) publication, 2009-125 and 2007-123.
Includes index.
ISBN 978-1-61761-818-5 (hardcover)
1. Nanotechnology--Health aspects. 2. Nanostructured materials--Health aspects. 3. Nanostructured materials industry--Safety measures. I. Caggione, Dominick A. II. National Institute for Occupational Safety and Health. III. Approaches to safe nanotechnology. IV. Progress toward safe nanotechnology in the workplace.
T174.7.S265 2010
620'.5--dc22

2010033089

Published by Nova Science Publishers, Inc. † New York

CONTENTS

PREFACE

The National Institute for Occupational Safety and Health (NIOSH) is the Federal agency responsible for conducting research and making recommendations to prevent work-related injury, illness, and death. As such, NIOSH is active in (1) identifying critical issues related to possible health hazards of nanomaterials, (2) protecting the safety and health of workers involved in this emerging technology, and (3) implementing a strategic plan to develop and disseminate methods for safely advancing the technology through workplace controls and safe handling procedures, and (4) investigating the possible applications of nanotechnology to solve workplace safety and health issues. Because of their small size and large surface area, engineered nanoparticles may have chemical, physical, and biological properties distinctly different from larger particles of similar chemical composition. Those properties may include the ability to reach the gas exchange regions of the lung, travel from the lung throughout the body, penetrate dermal barriers, cross cell membranes, and interact at the molecular level. NIOSH is investigating all of these properties, as it would with any new technology or material in the workplace, to provide the necessary guidance to ensure a safe and healthy workplace

Chapter 1- Nanotechnology has the potential to dramatically improve the effectiveness of a number of existing consumer and industrial products and could have a substantial impact on the development of new products in all sectors, ranging from disease diagnosis and treatment to environmental remediation. Because of the broad range of possible nanotechnology applications, continued evaluation of the potential health risks associated with exposure to nanomaterials is essential to ensure their safe handling. Engineered nanoparticles are materials purposefully produced with at least one dimension between 1 and 100 nanometers. Nanoparticles* often exhibit unique physical and chemical properties that impart specific characteristics essential in making engineered materials, but little is known about what effect these properties may have on human health. Research has shown that the physicochemical characteristics of particles can influence their effects in biological systems. These characteristics include particle size, shape, surface area, charge, chemical properties, solubility, oxidant generation potential, and degree of agglomeration. Until the results from research studies can fully elucidate the characteristics of nanoparticles that may pose a health risk, precautionary measures are warranted.

---

* In an attempt at standardization of terminology, the International Organization for Standardization-Technical Committee 229 has used the term nanomaterial to describe engineered nanoparticles.

NIOSH has developed this document to provide an overview of what is known about the potential hazards of engineered nanoparticles and measures that can be taken to minimize workplace exposures. Following is a summary of findings and key recommendations.

Chapter 2- Occupational safety and health issues of nanomaterials are complex. Because engineered nanomaterials are small yet have a relatively large surface area, they may have chemical, physical, and biological properties distinctly different from larger particles of similar chemical composition. Those properties may affect the ability of the nanomaterials to reach the gas exchange regions of the lung, travel from the lung throughout the body, penetrate dermal barriers, cross cell membranes, and interact at the molecular level. The types of nanomaterials and the opportunities for work place exposure to them continue to grow rapidly. The challenge is to effectively address the safety and health issues of nanotechnology while helping society realize nanotechnology's far-reaching potential benefits.

The National Institute for Occupational Safety and Health (NIOSH) is the Federal agency responsible for conducting research and making recommendations to prevent work-related injury, illness, and death. NIOSH is mandated by the Occupational Safety and Health Act to determine whether materials in a workplace constitute any harm and to provide recommendations for preventing injury and illness.

In: Safe Nanotechnology and Progress in the Workplace... ISBN: 978-1-61761-818-5
Editor: Dominick A. Caggione

Chapter 1

APPROACHES TO SAFE NANOTECHNOLOGY: MANAGING THE HEALTH AND SAFETY CONCERNS ASSOCIATED WITH ENGINEERED NANOMATERIALS*

Department of Health and Human Services

FOREWORD

Nanotechnology—the manipulation of matter on a near-atomic scale to produce new structures, materials, and devices—offers the promise of unprecedented scientific advancement for many sectors, such as medicine, consumer products, energy, materials, and manufacturing. Nanotechnology has the power not only to improve existing technologies, but to dramatically enhance the effectiveness of new applications.

Research on the potential applications of nanotechnology continues to expand rapidly worldwide. New nanotechnology consumer products emerge at a rate of three to four per week. Over the course of the next decade, nanotechnology could have a $1 trillion impact on the global economy and employ two million workers—half of them residing in the U.S.

While nanomaterials present seemingly limitless possibilities, they bring with them new challenges to understanding, predicting, and managing potential safety and health risks to workers. The National Institute for Occupational Safety and Health (NIOSH) remains committed to protecting workers now and in the future, as nanotechnology applications and uses expand.

As part of these efforts, in October 2005, NIOSH released for public comment the draft document, Approaches to Safe Nanotechnology: An Information Exchange with NIOSH. Based on feedback received, NIOSH revised and updated the document in July 2006 and sought further public comment. This draft report has been widely cited, and the final version of the report should serve as a vital resource for stakeholders (including occupational safety

*This is an edited, reformatted and augmented edition of a United States Department of Health and Human Services publication, Report 2009-125, dated March 2009.

and health professionals, researchers, policy makers, risk assessors, and workers in the industry) who wish to understand more about the safety and health implications of nanotechnology in the workplace.

With the publication of the Approaches to Safe Nanotechnology document, NIOSH hopes to: raise awareness of the occupational safety and health issues involved with nanotechnology; make recommendations on occupational safety and health best practices in the production and use of nanomaterials; facilitate dialogue between NIOSH and its external partners in industry, labor and academia; respond to requests for authoritative safety and health guidelines; and, identify information gaps and areas for future study and research.

As our knowledge of nanoscience increases, so too will our efforts to provide valuable guidance on the safe handling of nanoparticles and for protecting the lives and livelihoods of nanotechnology workers.

Christine M. Branche, Ph.D.
Acting Director, National Institute
for Occupational Safety and Health
Centers for Disease Control and Prevention

EXECUTIVE SUMMARY

Nanotechnology has the potential to dramatically improve the effectiveness of a number of existing consumer and industrial products and could have a substantial impact on the development of new products in all sectors, ranging from disease diagnosis and treatment to environmental remediation. Because of the broad range of possible nanotechnology applications, continued evaluation of the potential health risks associated with exposure to nanomaterials is essential to ensure their safe handling. Engineered nanoparticles are materials purposefully produced with at least one dimension between 1 and 100 nanometers. Nanoparticles* often exhibit unique physical and chemical properties that impart specific characteristics essential in making engineered materials, but little is known about what effect these properties may have on human health. Research has shown that the physicochemical characteristics of particles can influence their effects in biological systems. These characteristics include particle size, shape, surface area, charge, chemical properties, solubility, oxidant generation potential, and degree of agglomeration. Until the results from research studies can fully elucidate the characteristics of nanoparticles that may pose a health risk, precautionary measures are warranted.

* In an attempt at standardization of terminology, the International Organization for Standardization-Technical Committee 229 has used the term nanomaterial to describe engineered nanoparticles.

NIOSH has developed this document toprovide an overview of what is known about the potential hazards of engineerednanoparticles and measures that can betaken to minimize workplace exposures. Following is a summary of findings and key recommendations.

Potential Health Concerns

- The potential for nanomaterials to enter the body is among several factors that scientists examine in determining whether such materials may pose an occupational health hazard. Nanomaterials have the greatest potential to enter the body through the respiratory system if they are airborne and in the form of respirable-sized particles (nanoparticles). They may also come into contact with the skin or be ingested.
- Based on results from human and animal studies, airborne nanoparticles can be inhaled and deposit in the respiratory tract; and based on animal studies, nanoparticles can enter the blood stream, and translocate to other organs.
- Experimental studies in rats have shown that equivalent mass doses of insoluble incidental nanoparticles are more potent than large particles of similar composition in causing pulmonary inflammation and lung tumors. Results from in vitro cell culture studies with similar materials are generally supportive of the biological responses observed in animals.
- Experimental studies in animals, cell cultures, and cell-free systems have shown that changes in the chemicalcomposition, crystal structure, and sizeof particles can influence their oxidantgeneration properties and cytotoxicity.
- Studies in workers exposed to aerosols of some manufactured or incidental microscopic (fine) and nanoscale (ultrafine) particles have reported adverse lung effects including lung function decrements and obstructive and fibrotic lung diseases. The implications of these studies to engineered nanoparticles, which may have different particle properties, are uncertain.
- Research is needed to determine the key physical and chemical characteristics of nanoparticles that determine their hazard potential.

Potential Safety Concerns

- Although insufficient information exists to predict the fire and explosion risk associated with powders of nano- materials, nanoscale combustible material could present a higher risk than coarser material with a similar mass concentration given its increased particle surface area and potentially unique properties due to the nanoscale.
- Some nanomaterials may initiate catalytic reactions depending on their composition and structure that would not otherwise be anticipated based on their chemical composition.

Working with Engineered Nanomaterials

- Nanomaterial-enabled products such as nanocomposites, surface-coated materials, and materials comprised of nano- structures, such as integrated circuits,are unlikely to pose a risk of exposure during their handling and use as materials of non-inhalable size. However, some of the processes used in their production (e.g., formulating and applying nanoscale coatings) may lead to exposure to nanomaterials, and the cutting or grinding of such products could release respirable-sized nanoparticles.
- Maintenance on production systems (including cleaning and disposal of materials from dust collection systems) is likely to result in exposure to nanoparticles if deposited nanomaterials are disturbed.
- The following workplace tasks can increase the risk of exposure to nanoparticles:
    - Working with nanomaterials in liquid media without adequate protection (e.g., gloves)
    - Working with nanomaterials in liquid during pouring or mixing operations, or where a high degree of agitation is involved
    - Generating nanoparticles in non- enclosed systems
    - Handling (e.g., weighing, blending, spraying) powders of nano- materials
    - Maintenance on equipment and processes used to produce or fabricate nanomaterials and the cleaning-up of spills and waste material containing nanomaterials
    - Cleaning of dust collection systemsused to capture nanoparticles
    - Machining, sanding, drilling, or other mechanical disruptions of materials containing nanoparticles

Exposure Assessment and Characterization

- Until more information becomes available on the mechanisms underlying nanomaterial toxicity, it is uncertain what measurement technique should be used to monitor exposures in the workplace. Current research indicates that mass and bulk chemistry may be less important than particle size and shape, surface area, and surface chemistry (or activity) for some nanostructured materials.
- Many of the sampling techniques that are available for measuring airborne nanoaerosols vary in complexity but can provide useful information for evaluating occupational exposures with respect to particle size, mass, surface area, number concentration, and composition. Unfortunately, relatively few of these techniques are readily applicable to routine exposure monitoring. NIOSH has initiated exposure assessment studies in workplaces that manufacture or use engineered nanoparticles (see Appendix *Nanoparticle Emission Assessment Technique for Identification of Sources and Releases of Engineered Nanomaterials*).

- Regardless of the metric or measurement method used for evaluating nanoaerosol exposures, it is critical that background nanoscale particle measurements be conducted before the production, processing, or handling of nanomaterials.
- When feasible, personal sampling is preferred to ensure an accurate representation of the worker's exposure, whereas area sampling (e.g., size-fractionated aerosol samples) and real-time (direct reading) exposure measurements may be more useful for evaluating the needfor improvement of engineering controls and work practices.

Precautionary Measures

- Given the limited amount of information about health risks that may be associated with nanomaterials, taking measures to minimize worker exposures is prudent.
- For most processes and job tasks, the control of airborne exposure to nano- aerosols can be accomplished using a variety of engineering control techniques similar to those used in reducing exposure to general aerosols.
- The implementation of a risk management program in workplaces where exposure to nanomaterials exists can help to minimize the potential for exposure to nanoparticles. Elements of such a program should include the following:
    - Evaluating the hazard posed by the nanomaterial based on available physical and chemical property data, toxicology, or health-effects data
    - Assessing the worker's job task to determine the potential for exposure
    - Educating and training workers in the proper handling of nanomaterials (e.g., good work practices)
    - Establishing criteria and procedures for installing and evaluating engineering controls (e.g., exhaust ventilation) at locations where exposure to nanomaterials might occur
    - Developing procedures for determining the need for and selecting proper personal protective equipment (e.g., clothing, gloves, respirators)
    - Systematically evaluating exposures to ensure that control measures are working properly and that workers are being provided the appropriate personal protective equipment
- Engineering control techniques such as source enclosure (i.e., isolating the generation source from the worker) and local exhaust ventilation systems should be effective for capturing airborne nano- particles. Current knowledge indicates that a well-designed exhaust ventilation system with a high-efficiency particulate air (HEPA) filter should effectively remove nanomaterials.
- The use of good work practices can help to minimize worker exposures to nanomaterials. Examples of good practices include cleaning of work areas using HEPA vacuum pickup and wet wiping methods, preventing the consumption of food or beverages in workplaces where nanomaterials are handled, providing hand-washing facilities, and providing facilities for showering and changing clothes.
- No guidelines are currently available on the selection of clothing or other apparel (e.g., gloves) for the prevention of dermal exposure to nanoaerosols. However, some

clothing standards incorporate testing with nanometer-sized particles and therefore provide some indication of the effectiveness of protective clothing.

- Respirators may be necessary when engineering and administrative controls do not adequately prevent exposures. Currently, there are no specific limits for airborne exposures to engineered nanoparticles although occupational exposure limits exist for some larger particles of similar chemical composition. It should be recognized that exposure limits recommended for nonnanoscale particles may not be health protective for nanoparticle exposures (e.g., the OSHA Permissible Exposure Limit [PEL] for graphite may not be a safe exposure limit for carbon nano- tubes). The decision to use respiratory protection should be based on professional judgment that takes into account toxicity information, exposure measurement data, and the frequency and likelihood of the worker's exposure. While research is continuing, preliminary evidence indicates that NIOSH-certified respirators will be useful for protecting workers from nanoparticle inhalation when properly selected and fit tested as part of a complete respiratory protection program.

Occupational Health Surveillance

Occupational health surveillance is an essential component of an effective occupational safety and health program. The unique physical and chemical properties of nanomaterials, the increasing growth of nanotechnology in the workplace, and information suggesting that exposure to some engineered nanomaterials can cause adverse health effects in laboratory animals all support consideration of an occupational health surveillance program for workers potentially exposed to engineerednanomaterials. Continued evaluation of toxicologic research and workers potentially exposed to engineered nanomaterials is needed to inform NIOSH and other groups regarding the appropriate components of occupational health surveillance for nanotechnology workers. NIOSH has formulated interim guidance relevant to medical screening (one component of an occupational health surveillance program) for nanotechnology workers (see NIOSH *Current Intelligence Bulletin InterimGuidance for Medical Screening and Hazard Surveillance for Workers Potentially Exposed to Engineered Nanoparticles* at www.cdc.gov/ niosh/review/public/ 115/). In this document NIOSH concluded that insufficient scientific and medical evidence now exist to recommend the specific medical screening of workers potentially exposed to engineered nano- particles. However, NIOSH did recommend that hazard surveillance be conducted as the basis for implementing control measures.

ACKNOWLEDGMENTS

This chapter was developed by the scientists and staff of the National Institute for Occupational Safety and Health (NIOSH) who participate in the NIOSH Nanotechnology Research Center (NTRC). Paul Schulte is the manager and Charles Geraci, coordinator of the NIOSH NORA nanotechnology cross-sector program. Special thanks go to Ralph Zumwalde and Laura Hodson for writing and organizing the report and to Mark Methner for the development of the *Appendix: Nanoparticle Emission Assessment Technique for Identification of Sources and Releases of Engineered Nanomaterials*.

Others who contributed substantially to the writing and research described here include: Eileen Birch, Fred Blosser, Vincent Castranova, Brian Curwin, Douglas Evans, Pengfei Gao, Donna Heidel, Mark Hoover, John Howard, Vijia Karra, Bon-Ki Ku, Eileen Kuempel, Robert Mercer, Arthur Miller, Vladimir Murashov, Terri Pearce, Appavoo Rengasamy, Ronald Shaffer, Anna Shvedova, Petia Simeonova, Aleksandr Stefaniak, Douglas Trout, and Leonid Turkevich.

The NIOSH NTRC also acknowledges the contributions of Vanessa Becks and Gino Fazio for desktop publishing and graphic design, and Elizabeth Fryer for editing the report.

1. INTRODUCTION

Nanotechnology is the manipulation of matter on a near-atomic scale to produce new structures, materials, and devices. This technology has the ability to transform many industries and can be applied in many ways to areas ranging from medicine to manufacturing. Research in nanoscale technologies is growing rapidly worldwide. Lux Research [2007] projects that new emerging nanotechnology applications will affect nearly every type of manufactured product through the middle of the next decade, becoming incorporated into 15% of global manufacturing output, totaling $2.6 trillion in 2014.

Nanomaterials present new challenges to understanding, predicting, and managing potential health risks to workers. As with any material being developed, scientific data on the health effects in exposed workers are largely unavailable. In the case of nanomaterials, the uncertainties are great because the characteristics of nanoparticles may be different from those of larger particles with the same chemical composition. Safety and health practitioners recognize the critical lack of specific guidance on the safe handling of nanomaterials—especially now, when the degree of risk to exposed workersis unknown. In the meantime, the extensive scientific literature on airborne particles— including toxicology and epidemiological studies, measurement techniques, and engineering controls—provides the best available data from which to develop interim approaches for working safely with nanomaterials and to develop hypotheses for studies of new nanomaterials.

The National Institute for Occupational Safety and Health (NIOSH) is working in parallel with the development and implementation of commercial nanotechnology through (*1*) conducting strategic planning and research, (*2*) partnering with public- and private-sector colleagues from the United States and abroad, and (*3*), making information widely available. The NIOSH goal is to provide national and world leadership for incorporating research findings about the implications and applications of nanotechnology into good occupational safety and health practice for the benefit of all nano- technology workers. NIOSH has developed a strategic plan for coordinating nanotechnology research and for use as a guide for enhancing the development of new research efforts(www.cdc.gov/niosh/topics/nanotech/strat_plan.html).

2. Purpose

With the publication of this *Approaches to Safe Nanotechnology* document, NIOSH hopes to do the following:

- **Raise awareness** of the occupational safety and health issues being identified in the rapidly moving and changing science involving implications and applications of nanotechnology.
- **Use the best information available to make recommendations** on occupational safety and health practices in the production and use of nanomaterials (These recommendations will be updated as appropriate to reflect new information. They will address key components of occupational safety and health, including exposure monitoring, engineering controls, personal protective equipment, and administrative controls. They willdraw from the ongoing NIOSH assessment of current best practices, technical knowledge, and professional judgment. Throughout the development of these guidelines, the utility of a hazard-based approach to risk assessment and control was evaluated and, where appropriate, recommendations are provided.)
- **Facilitate an exchange of information** between NIOSH and its external partners from ongoing research, including success stories, applications, and case studies.
- **Respond to requests** from industry, labor, academia, and other partners who are seeking science-based, authoritative guidelines.
- **Identify information gaps** where few or no data exist and where research is needed.

3. Scope

This document has been developed to provide a resource for stakeholders who wish to understand more about the safety and health implications and applications of nanotechnology in the workplace. The information and guidelines presented here are intended to aid in evaluating the potential hazard of exposure to engineered nanomaterials and to set the stage for the development of more comprehensive guidelines for reducing potential workplace exposures in the wide range of tasks and processes that use nanomaterials. The information in this document will be of specific interest to the following:

- Occupational safety and health professionals who must (*1*) understand how nanotechnology may affect occupational health and (*2*) devise strategies for working safely with nanomaterials
- Researchers working with or planning to work with engineered nanomaterials and studying the potential occupational safety and health impacts of nanomaterials
- Policy and decision makers in government agencies and industry
- Risk evaluation professionals
- People working with or potentially exposed to engineered nanomaterials in the workplace

Established safe work practices are generally based on an understanding of the hazards associated with the chemical and physical properties of a material. Engineered nanomaterials may exhibit unique properties that are related to their physical size, shape, structure, and chemical composition. Considerable uncertainty still exists as to whether these unique properties present occupational health risks. Current information about the potential adverse health effects of engineered nanomaterials, exposure assessment, and exposure control is limited. However, the large body of scientific literature that exists on exposures to and responses of animals and humans to ultrafine and other airborne particles may be useful in making preliminary assessments as to the health risks posed by engineered nanomaterials. Until further information is available, interim safe working practices should be usedbased on the best available information.The information and recommendations in this document are intended to aid in assessment of the potential hazard of engineered nanomaterials and to set the stage for the development of more comprehensive guidelines for reducing potential workplace exposures.

4. Descriptions and Definitions

Nanotechnology involves the manipulation of matter at nanometer[†] scales to produce new materials, structures, and devices. The U.S. National Nanotechnology Initiative (see http://nano.gov/html/facts/whatIsNano. html) defines a technology as nanotechnology only if it involves all of the following:

- Research and technology development involving structures with at least one dimension in the range of 1–100 nanometers (nm), frequently with atomic/ molecular precision
- Creating and using structures, devices, and systems that have unique properties and functions because of their nanoscale dimensions
- The ability to control or manipulate on the atomic scale

Nanotechnology is an enabling technology that offers the potential for unprecedented advances in many diverse fields. The ability to manipulate matter at the atomic or molecular scale makes it possible to form new materials, structures, and devices that exploit the unique physical and chemical properties associated with nanoscale structures. The promise of nanotechnology goes far beyond extending the use of current materials. New materials and devices with intricate and closely engineered structures will allow for (*1*) new directions in optics, electronics, and optoelectronics, (*2*) development of new medical imaging and treatment technologies, and (*3*) production ofadvanced materials with unique properties and high-efficiency energy storage and generation.

Although nanotechnology-based products are generally thought to be at the precompetitive stage, an increasing number of products and materials are becoming commercially available. These include nanoscale powders, solutions, and suspensions of nanoscale materials as well as composite materials and devices having a nanostructure. Nanoscale products

[†] 1 nanometer (nm) = 1 billionth of a meter (10^{-9}).

and materials are increasingly used in optoelectronic, electronic, magnetic, medical imaging, drug delivery, cosmetic, catalytic, and materials applications. New nano-technology consumer products are coming on the market at the rate of three to four per week, a finding based on the latest update to the nanotechnology consumer product inventory maintained by the Project on Emerging Nanotechnologies (PEN)[‡] (www. nanotechproject.org/ inventories/consumer). The number of consumer products using nanotechnology has grown from 212 to 609 since PEN launched the world's first online inventory of manufacturer-identified nano- tech goods in March 2006.

Figure 4–1. Photomicrographs of airborne exposure to ultrafine (nanoscale) particles of welding fumes, diesel exhaust, and cerium oxide

According to Lux Research [2007], in 2006, governments, corporations, and venture capitalists worldwide spent $11.8 billion on nano- technology research and development (R&D), which was up 13% from 2005. By 2014, Lux estimates $2.6 trillion in manufactured goods will incorporate nanotechnology—or about 15% of total global output.

4.1. Nano-objects

The International Organization for Standardization Technical Committee 229 (Nanotechnologies) is developing globally recognized nomenclature and terminology for nanomaterials. According to ISO/TS 27687:2008, *nano-object* is defined as material with one, two, or three external dimensions in the size range from approximately 1–100 nm. Subcategories of nano-object are (*1*) *nanoplate*, a nano-object with one external dimension at the nanoscale; (*2*) *nanofiber*, a nano-object with two external dimensions at the nano- scale with a nanotube defined as a hollow nanofiber and a nanorod as a solid nanofiber; and (*3*) *nanoparticle*, a nano-object with all three external dimensions at the nanoscale. Nano-objects are commonly incorporated in a larger matrix or substrate referred to as a *nanomaterial*. Nano-objects may be suspended in a gas (as a nanoaerosol), suspended in a liquid (as a colloid or nanohydrosol), or embedded in a matrix (as a nanocomposite).

The precise definition of particle diameterdepends on particle shape as well as how the diameter is measured. Particle morphologiesmay vary widely at the nanoscale. For instance, carbon fullerenes represent nano- objects with identical dimensions in all directions (i.e., spherical), whereas single -walled carbon nanotubes (SWCNTs) typically form convoluted, fiber-like nano-objects. Many regular but nonspherical particle morphologies can be engineered at the nanoscale, including flower- and belt-like structures. Please see

[‡] The Project on Emerging Nanotechnologies was established in April 2005 as a partnership between the Woodrow Wilson International Center for Scholars and the Pew Charitable Trusts.

www.nanoscience.gatech.edu/zlwang/research.html for examples of some nanoscale structures.

4.2. Ultrafine Particles

The term *ultrafine particle* has traditionally been used by the aerosol research and occupational and environmental health communities to describe airborne particles smaller than 100 nm in diameter. *Ultrafine* is frequently used in the context of nanometer-diameter particles that have not been intentionally produced but are the incidental productsof processes involving combustion, welding, or diesel engines (see Figure 4–1). The term *nanoparticle* is frequently used with respect to particles demonstrating size-dependent physicochemical properties, particularly from a materials science perspective. The two terms are sometimes used to differentiate between engineered(*nanoparticle*) and incidental (*ultrafine*) nanoscale particles.

It is currently unclear whether the use of source-based definitions of nanoparticles and ultrafine particles is justified from a safety and health perspective. This is particularly the case where data on non- engineered, nanometer-diameter particles are of direct relevance to the impact of engineered particles. An attempt has been made in this document to follow the general convention of preferentially using *nanoparticle* in the context of intentionally produced or engineered particles and *ultrafine* in the context of incidentally produced particles (e.g., combustion products). However, this does not necessarily imply specific differences in the properties of these particles as related to hazard assessment, measurement, or control of exposures, and this remains an active area of research. *Nanoparticle* and *ultrafine particle* are not rigid definitions. For example, since the term *ultrafine* has been in existence longer, some intentionally produced particles with primary particle sizes in the nanosize range (e.g., TiO_2) are often called *ultrafine* in the literature.

4.3. Engineered Nanoparticles

Engineered nanoparticles are intentionally produced, whereas *ultrafine particles* (often referred to as *incidental nanoparticles)* are typically byproducts of processes such as combustion and vaporization. Engineered nanoparticles are designed with very specific properties or compositions (e.g., shape, size, surface properties, and chemistry). Incidental nanoparticles are generated in a relatively uncontrolled manner and are usuallyphysically and chemically heterogeneous compared with engineered nanoparticles.

4.4. Nanoaerosol

A *nanoaerosol* is a collection of nanoparticles suspended in a gas. The particles may be present as discrete nano-objects, or as aggregates or agglomerates of nano- objects. These agglomerates may have diameters larger than 100 nm. In the case of an aerosol consisting of micrometer- diameter particles formed as agglomerates of nano-objects, the definition of nanoaerosol is open to interpretation. It is generally accepted that if the nanostructure associated with the nano- object is accessible (through physical, chemical, or biological interactions), then the aerosol may be considered a nanoaerosol. However, if the nanostructure within individual micrometer-diameter particles does not directly influence particle behavior (for instance, if the nanoparticles were inaccessibly embedded in a solid matrix), the aerosol would not be described as a nanoaerosol.

4.5. Agglomerate

An *agglomerate* is a group of nanoparticles held together by relatively weak forces, in including van der Waals forces, electrostatic forces, and surface tension [ISO 2006].

4.6. Aggregate

An *aggregate* is a heterogeneous particle in which the various components are held together by relatively strong forces, and thus not easily broken apart [ISO 2006]. Aggregated nanoparticles would be an example of a nanostructured material.

5. Potential Health Concerns

Nanotechnology is an emerging field. As such, there are many uncertainties as to whether the unique properties of engineered nanomaterials (which underpin their commercial and scientific potential) also pose occupational health risks. These uncertainties arise because of gaps in knowledge about the factors that are essential for predicting health risks—factors such as routes of exposure, translocation of materials once they enter the body, and interaction of the materials with the body's biological systems. The potential health risk following exposure to a substance is generally associated with the magnitude and duration of the exposure, the persistence of the material in the body, the inherent toxicity of the material, and the susceptibility or health status of the person exposed. More data are needed on the health risks associated with exposure to engineered nanomaterials. Results of existing studies in animals and humans on exposure and response to ultra- fine or other respirable particles provide a basis for preliminary estimates of the possible adverse health effects from exposures to similar engineered materials on a nano- scale. Experimental studies in rodents and cell cultures have shown that the toxicity of ultrafine or nanoparticles is greater than that of the same mass of larger particles of similar chemical composition [Oberdörster et al. 1992, 1994a, b; Lison et al. 1997; Tran et al. 1999, 2000; Brown et al. 2001; Barlow et al. 2005; Duffin et al. 2007]. In addition to particle surface area, other particle characteristics may influence toxicity, including surface functionalization or coatings, solubility, shape, and the ability to generateoxidant species and to adsorb biological proteins or bind to receptors [Duffin et al. 2002; Oberdörster et al. 2005a; Maynard and Kuempel 2005; Donaldson et al. 2006]. More research is needed on the influence of particle properties on interactions with biological systems and the potential for adverse effects. International research strategies for evaluating the safety of nanomaterials are actively being developed through cooperative efforts [Thomas et al. 2006].

Existing toxicity information about a given material of larger particle size can provide a baseline for anticipating the possible adverse health effects that may occur from exposure to a nanoscale material that has some of the same physicochemical properties (e.g., chemistry, density). However, predicting the toxicity of an engineered nanomaterial based on its physicochemical properties may not provide an adequate level of protection.

5.1. Exposure Routes

Inhalation is the most common route of exposure to airborne particles in the workplace. The deposition of discrete nano- objects in the respiratory tract is determined by the particle's aerodynamic or thermodynamic diameter (i.e., the particle shape and size). Agglomerates of nano-objects will deposit according to the diameter of the agglomerate, not constituent nano-objects. Research is ongoing to determine the physical factors that contribute to the agglomeration and de-agglomeration of nano-objects in air, suspended in aqueous media, or once in contact with lung lining fluid and/or biological proteins. Evidence indicatesthat the degree of agglomeration can affect the toxicity of inhaled nano-objects [Shvedova et al. 2007].

Discrete nanoparticles are deposited in the lungs to a greater extent than larger respirable particles [ICRP 1994], and deposition increases with exercise due to increase in breathing rate and change from nasal to mouth breathing [Jaques and Kim 2000; Daigle et al. 2003] and among persons with existing lung diseases or conditions (e.g., asthma, emphysema) [Brown et al. 2002]. Based on animal studies, discrete nano- particles may enter the bloodstream from the lungs and translocate to other organs [Takenaka et al. 2001; Nemmar et al. 2002; Oberdörster et al. 2002].

Discrete nanoparticles (35–37-nm median diameter) that deposit in the nasal region may be able to enter the brain by translocation along the olfactory nerve, as was observed in rats [Oberdörster et al. 2004; Oberdörster et al. 2005a; Elder et al. 2006]. The transport of insoluble particles from 20–500 nm-diameter to the brain via sensory nerves (including olfactory and trigeminus) was reported in earlier studies in several animal models [De Lorenzo 1970; Adams and Bray 1983; Hunter and Dey 1998]. This exposure route for nanoparticles and to nanoscale biological agents has not been studied in humans.

Some studies suggest that nanomaterials could potentially enter the body through the skin during occupational exposure. Tinkle et al. [2003] have shown that particles smaller than 1 μm in diameter may penetrate into mechanically flexed skin samples. A more recent study reported that nanoparticles with varying physicochemical properties were able to penetrate the intact skin of pigs [Ryman-Rasmussen et al. 2006]. Thesenanoparticles were quantum dots of different size, shape, and surface coatings. They were reported to penetrate the stratum corenum barrier by passive diffusion and localize within the epidermal and dermal layers within 8–24 hours. The dosing solutions were 2- to 4-fold dilutions of quantum dots as commercially supplied and thus represent occupationally relevant doses.

At this time, it is not fully known whether skin penetration of nanoparticles would result in adverse effects in animal models. However, topical application of raw SWCNT to nude mice has been shown to cause dermal irritation [Murray et al. 2007]. Studies conducted in vitro using primary or cultured human skin cells have shown that both SWCNT and multi-walled carbon nanotubes (MWCNT) can enter cells and cause release of pro-inflammatory cytokines, oxidative stress, and decreased viability [Monteiro-Riviere et al. 2005; Shvedova et al. 2003]. It remains unclear, however, how these findings may be extrapolated to a potential occupational risk, given that additional data are not yet available for comparing the cell model studies with actual conditions of occupational exposure. Research on the dermal exposure of nanomaterials is ongoing (www.unileipzig.de/~nanoderm/).

Ingestion can occur from unintentional hand to mouth transfer of materials; this has been found to happen with traditional materials, and it is scientifically reasonable to assume that it also could happen during handling of nanomaterials. Ingestion may also accompany inhalation exposure because particles that are cleared from the respiratorytract via the mucociliary escalator may be swallowed [ICRP 1994]. Little is known about possible adverse effects from the ingestion of nanomaterials.

5.2. Effects Seen in Animal Studies

Experimental studies in rats have shown that at equivalent mass doses, insoluble ultrafine particles are more potent than larger particles of similar composition in causing pulmonary inflammation, tissue damage, and lung tumors[Lee et al. 1985; Oberdörster and Yu 1990; Oberdörster et al. 1992, 1994a,b; Heinrich et al. 1995; Driscoll 1996; Lison et al. 1997; Tran et al. 1999, 2000; Brown et al. 2001; Duffin et al. 2002; Renwick et al. 2004; Barlow et al. 2005]. These studies have shown that for poorly-soluble low toxicity (PSLT) particles, the dose-response relationships are consistent across particle sizes when dose is expressed as particle surface area. In addition to particle size and surface area, studies have shown that other particle characteristics can influence toxicity. For example, although the relationship betweenparticle surface area dose and pulmonary inflammation is consistent among PSLT particles, crystalline silica is much more inflammogenic than PSLT particles at a given surface area dose [Duffin et al. 2007].

Reactive oxidant generation on the particle surface is an important factor influencing lung response to particles, which can be related to crystal structure. A recent study of the lung effects of rats dosed with either ultrafine *anatase* titanium dioxide (TiO_2) or ultrafine *rutile* TiO_2 showed that the *anatase* TiO_2 had more reactive surfaces and caused greater pulmonary inflammation and cell proliferation in the lungs of rats [Warheit et al. 2007]. In a cell-free assay designed to investigate the role of surface area and crystal structure on particle reactive oxygen species (ROS)-generation, Jiang et al. [2008] observed that size, surface area, and crystal structure all contribute to ROS generation.

Oxidant generation was apparently associated with the number of defective sites per surface area, which varied in nanoparticles in some size ranges [Jiang et al. 2008].

These studies indicate that for nanoparticles with similar properties (e.g., PSLT), the toxicity of a given mass dose will increase with decreasing particle size due to the increasing surface area. However, the dose-response relationship may differ for particles with different chemical composition and other properties. Consistent with these findings, a recent pulmonary instillation study with rats dosed with either fine or ultrafine TiO_2 reported no significant difference in lung responses when compared to controls, while crystalline silica caused more severe lung responses at the same dose [Warheit et al. 2006]. However, Warheit et al. [2006] were unable to adequately test the hypotheses about the relationship between particle surface area dose and toxicity because the diameters of the fine and ultrafine TiO_2-instilled particles did not significantly differ due to particle agglomeration, both being in excess of 2 μm. When efforts were made to more effectively disperse fine and ultrafine particles, the effect of surface area on the pulmonary response in rats after intratracheal instillation was verified [Sager et al. 2008].

5.2.1. Polytetrafluoroethylene fume

Among ultrafine particles, freshly generated polytetrafluoroethylene (PTFE) fume (generated at temperatures of more than 425°C) is known to be highly toxic to the lungs. Freshly generated PTFE fume caused hemorrhagic pulmonary edema and death in rats exposed to less than 60 $\mu g/m^3$ [Oberdörster et al. 1995]. In contrast, aged PTFE fume was much less toxic and did not result in mortality. This low toxicity was attributed to the increase in particle size from accumulation and to changes in surface chemistry [Johnston et al. 2000; Oberdörster et al. 2005a]. Human case studies have reported pulmonary edema in workers exposed to PTFE fume and an accidental death in a worker when an equipment malfunction caused overheating of the PTFE resin and release of the PTFE pyrolysis products in the workplace [Goldstein et al. 1987; Lee et al. 1997]. While PTFE fume differs from engineered nanoparticles, these studies illustrate properties of ultrafine particles that have been associated with an acute toxic hazard. Enclosed processes and other engineering controls appear to have been effective at eliminating worker exposures to PTFE fume in normal operations, and thus may provide examples of control systems that may be implemented to prevent exposure to nanoparticles that may have similar properties.

5.2.2. Carbon nanotubes

Carbon nanotubes (CNT) are specialized forms or structures of engineered nanomaterials that have had increasing production and use [Donaldson et al. 2006]. Consequently, a number of toxicologic studies of CNT have been performed in recent years. These studies have shown that the toxicity of CNT may differ from that of other nanomaterials of similar chemical composition. For example, single-walled CNTs (SWCNT) have been shown to produce adverse effects including granulomas in the lungs of mice and rats at mass doses at which ultrafine carbon black did not produce these adverse effects [Shvedova et al. 2005; Lam et al. 2004]. While both SWCNTs and carbon black are carbon-based, SWCNTs have a unique, convoluted, fibrous structure and specific surface chemistry that offers excellent electrical conductive properties. How these characteristics may influence toxicity is not known. Carbon nanotubes may contain metal catalysts as byproducts of their production, which could contribute to their toxicity, or the CNTs may provide a structure that promotes fibroblast cell growth [Wang et al. 2008].

In a study of SWCNTs instilled into the lungs of rats, multi-focal granulomas (without transient inflammation or persistent lesions) were observed at doses of 1 or 5 mg / kg body weight [Warheit et al. 2004]. In a study of mice instilled with one of several types of SWCNTs (i.e., raw, purified, iron-containing, and nickel-containing) at doses of 0.1 or 0.5 mg/mouse (approximately 3 or 16 mg/kg body weight), dose-dependent epithelioid granulomas were observed at 7 days, which persisted at 90 days [Lam et al. 2004, 2006]. Both the raw and purified forms produced interstitial inflammation, while mortality (5/9 mice) was observed in the high dose group of the Ni-containing SWCNT.

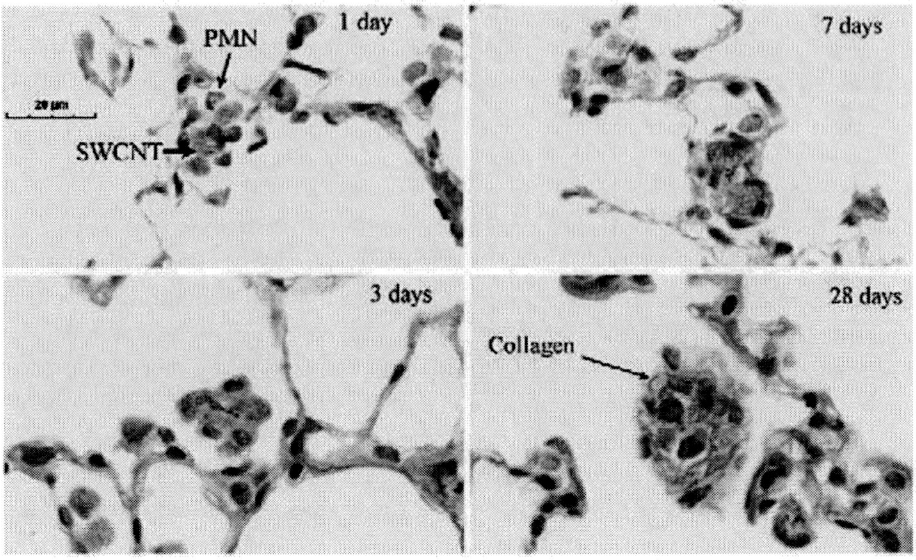

Figure 5–1. Formation of collagen following deposition of SWCNTs in the lungs of mice

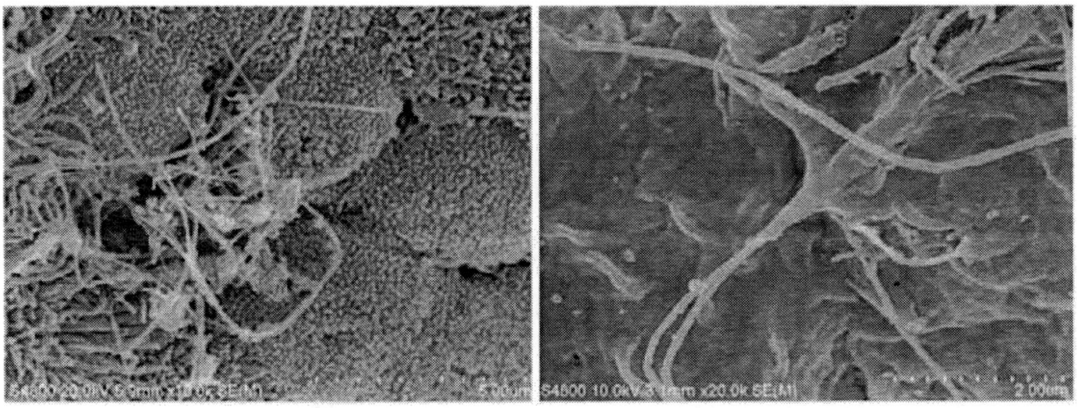

Figure 5–2. Deposition and clearance of MWCNTs from the conducting airways of mice following inhalation exposure

NIOSH researchers recently reported adverse lung effects following pharyngeal aspiration of SWCNTs in mice using doses between 10–40 μg/mouse (approximately 0.5–2 mg / kgbody weight) [Shvedova et al. 2005]. The findings showed that exposure to SWCNTs in mice lead to transient pulmonary inflammation, oxidative stress, decrease in pulmonary function, decrease in bacterial clearance, and early onset of interstitial fibrosis. Deposition of agglomerates resulted in development of granulomas, while deposition of dispersed nanotube structures in the aspirated suspension resulted in the rapid development of interstitial fibrosis (within 7 days), which progressed over a 30–60 day post-exposure period [Shvedova et al. 2005; Mercer et al. 2008]. Evidence indicates that when efforts were made to more fully disperse the SWCNT and obtain smaller structures in the aspiration suspension, fewer granulomas occurred but a 4-fold more potent interstitial fibrotic response was observed [Mercer et al. 2008].

Exposure to SWCNT has been observed to be more fibrogenic than an equal mass of either ultrafine carbon black or fine quartz [Shvedova et al. 2005; Lam et al. 2004]. Based on their findings in mice, Shvedova et al. [2005] estimated that workers may be at risk of developing lung lesions if they were exposed to SWCNT over a period of 20 days at the currentOSHA PEL for graphite (5 mg/m^3). Lam et al. [2004, 2006] provided similar estimates and suggested that the graphite PEL should not be used (e.g., on MSDS) as a safe concentration for workers exposed to CNTs. Compared to instillation, the pharyngeal aspiration technique may approximate more closely the particle deposition that occurs during inhalation. Inhalation studies of CNTs may provide more definitive information about their potential toxicity in humans [Donaldson et al. 2006]. Recently, NIOSH scientists designed a system to generate an aerosol of SWCNT for a rat inhalation study [Baron et al. 2008]. Results of the inhalation exposure to SWCNT [Shvedova et al. 2008] were qualitatively similar to those of the aspiration study [Shvedova et al. 2004] with a 4-fold more potent interstitial fibrotic response similar to that reported by Mercer et al. [2008]. Another NIOSH study found markers of inflammation in the lung, aorta, and heart tissues of ApoE-/- mice after a single intra-pharyngeal instillation dose of SWCNT (10 and 40 μg/mouse) and accelerated plaque formation after repeated doses (20 μg/mouse once every other week for 8 weeks in mice fed an atherogenic diet) [Li et al. 2007].

MWCNTs were recently studied by intratracheal instillation in Sprague-Dawley rats receiving 0.5, 2, or 5 mg (approximately 2, 9, or 22 mg/kg body weight) of either ground MWCNT or unground MWCNT [Muller et al. 2005]. Both forms produced pulmonary inflammation and fibrosis. Rats that received ground MWCNT showed greater dispersion in the lungs, and fibrotic lesions were observed in the deep lungs (alveolar region). In rats treated with MWCNT (not ground) fibrosis showed mainly in their airways rather than in their lungs. The biopersistence of the unground MWCNT was greater than that of the ground MWCNT, with 81% vs. 36%, respectively, remaining in the lungs at day 60.

At an equal mass dose, ground MWCNT produced a similar inflammatory and fibrogenic response as chrysotile asbestos and a greater response than ultrafine carbon black [Muller et al. 2005]. Effects from the vehicle (1% Tween 80) used for administering ground and unground MWCNT to rats were not reported; the control group used in the study was exposed to only saline. NIOSH scientists have exposed mice by aspiration to MWCNT suspended in a simulated alveolar lining fluid rather than Tween 80. Control studies show that this suspension medium was not inflammatory and did not mask the biological activity of the particle surface. Data indicate that aspiration of dispersed MWCNT produced pulmonary inflammation, which peaked 7 days post exposure. The inflammatory response to MWCNT was greater than the inflammatory response to SWCNT [Sriram et al. 2007].

Two recent studies investigated the hypotheses that CNTs can behave like asbestos. In the first study, Takagi et al. [2008] administered to p53 (+/-) mice MWCNT, fullerene, or crocidolite asbestos by intraperitoneal injection at doses of 3 mg/mouse. The average width of the MWCNT was approximately 100 nm, and approximately 28% of the particles were longer than 5 μm. The particle number concentrations of MWCNT and crocidolite were 1×10^9 and 1×10^{10} (in 1-ml suspensions), respectively, although the MWCNT sample was also reported to contain mainly large aggregates, indicating that the number of MWCNT fibers was vastly underestimated and much larger than for the asbestos exposure. At the termination of the study (25 weeks), mesothelial responses in the MWCNTtreated mice included moderate to severe fibrous peritoneal adhesion and peritoneal tumors. The asbestos-treated mice had similar responses but to a lesser extent, while thefullerene-treated group did not show these responses. Mesothelioma was considered by the authors as the primary cause of death, and constriction of the ileus due to severe peritoneal adhesion was considered to be the second major cause of death, suggesting that 3 mg/mouse exceeded the maximum tolerated dose of MWCNT. Whether mesotheliomia was a primary cause of death is somewhat speculative.

In a second study, Poland et al. [2008] administered to mice either MWCNT (two short and two long CNT samples), nanoscale carbon black, or amosite (short or long) at doses of 50 μg/mouse by intraperitoneal injection. The short CNTs were 10 nm or 15 nm in width, with no fibers larger than 15 μm in length detected; the long CNTs were 85 nm or 165 nm in width, and 24% or 84%, respectively, were larger than 15 μm in length (the percentage of fibers longer than 5 μm was not reported). After either 24 hours or 7 days, the long MWCNT caused inflammation and granulomatous lesions that were qualitatively and quantitatively similar to that caused by the long asbestos. The short, low-aspect-ratio, tangled aggregates of MWCNT did not produce these responses at the doses used in this study. Additional studies are needed to determine if this inflammatory response to MWCNT would be persistent and result in tumors of the abdominal wall. Additionally, the potential for migration of MWCNT through the lungs to the mesothelium after inhalation requires investigation. Long-term studies are also needed to determine whether CNTs can cause cancer such as mesothelioma in laboratory animals, including exposures by typical routes in humans (i.e., inhalation, dermal penetration, and ingestion) and at doses that include those equivalent to potential workplace exposures.

These studies indicate the need for more data on exposures of workers to CNTs. Maynard et al. [2004] reported relatively low short-term (approximately 30 min) airborne mass concentrations of SWCNT (0.007–0.053 mg/m^3) in a laboratory production facility. A recent study by Han et al. [2008] reported total airborne mass concentrations of MWCNT from 0.21–0.43 mg/m^3 (4–6-hr sampling) in a laboratory research facility prior to use of engineering control measures; after implementing controls, the concentration decreased to nondetectable. Workers could also be exposed to ground CNTs used in polymer composites and other matrices or during cutting, grinding, or polishing of these materials. Given that exposure to SWCNT and MWCNT causes interstitial fibrosis and pulmonary inflammation, respectively, in rodent lungs at relatively low mass doses, it is prudent to minimize worker exposure to airborne CNTs (see Chapter 8 *Guidelines for Working with Engineered Nanomaterials*).

5.3. Observations from Epidemiological Studies Involving Fine and Ultrafine Particles

Epidemiological studies in workers exposed to aerosols including fine and ultrafine particles have reported lung function decrements, adverse respiratory symptoms, chronic obstructive pulmonary disease, and fibrosis [Kreiss et al. 1997; Gardiner et al. 2001; Antonini 2003]. In addition, some studies have found lung disease including elevated lung cancer and neurological effects among workers exposed to certain ultrafine particles (i.e., diesel exhaust particulate) [Steenland et al. 1998; Garshick et al. 2004, 2006; Hart et al. 2006] or weldingfumes [Antonini 2003; Parket al. 2006; Ambroise et al. 2007; Bowler et al. 2007]. The implications of these studies to engineered nanomaterials, which may have different particle properties, are uncertain. Studies of airborne particles and fibers in the workplace do provide relevant background information about the particle-related lung diseases and mechanisms, and some limited quantitative estimates of exposures and risk of adverse health effects. As such, these studies provide a point of reference, including baseline information and estimates regarding possible health risks of exposure to other nanoscale particles depending on the extent to which the exposure conditions and particle -biological interactions may be similar.

Epidemiological studies in the general population have also shown associations between particulate air pollution and increased morbidity and mortality from respiratory and cardiovascular diseases [Dockery et al. 1993; HEI 2000; Pope et al. 2002, 2004]. Some epidemiological studies have shown adverse health effects associated with exposure to the ultrafine particulate fraction of air pollution [Peters et al. 1997, 2004; Penttinen et al. 2001; Ibald-Mulli et al. 2002; Timonen et al. 2004; Ruckerl et al. 2006] although uncertainty exists about the role of ultra- fine particles relative to other air pollutants in causing the observed adverse health effects. The associations in these studies have been based on measurements of the particle number or mass concentrations of particles within certain size fractions (e.g., particulate matter with diameter of 2.5 μm and smaller [$PM_{2.5}$]). In an experimental study of healthy and asthmatic subjects inhaling ultrafine carbon particles, changes were observed in the expression of adhesion molecules by blood leukocyte, which may relate to possible cardiovascular effects of ultrafine particle exposure [Frampton et al. 2006]. Short-term diesel exhaust exposure (0.3 mg/ m^3 for 1 hr) in healthy volunteers was associated with mild systemic inflammation and impaired endothelial-dependent vasodilation [Törnqvist et al. 2007].

5.4. Hypotheses from Animaland EpidemiologicalStudies

The existing literature on particles and fibers provides a scientific basis from which to evaluate the potential hazards of engineered nanomaterials. While the properties of engineered nanomaterials can vary widely, the basic physicochemical and toxicokinetic principles learned from the existing studies are relevant to understanding the potential toxicity of nanomaterials. For example, it is known from studies in humans that a greater proportion of inhaled nanoparticles will deposit in the respiratory tract (both at rest and with exercise) compared to larger particles [ICRP 1994; Jaques and Kim 2000; Daigle et al. 2003; Kim and Jaques 2004]. It is also known from studies in animals that nanoparticles in the lungs can be trans- located to other organs in the body; how the chemical and physical properties of the nanoparticles influence this translocation is not completely known [Takenaka et al. 2001; Kreyling et al. 2002; Oberdörster et al. 2002, 2004; Semmler et al. 2004; Geiser et al. 2005]. Due to their small size, nanoparticles can cross cell membranes and interact with subcellular structures such as mitochondria, where they have been shown tocause oxidative damage and to impair function of cells in culture [Möller et al. 2002, 2005; Li et al. 2003; Geiser et al.

2005]. Nanoparticles have also been observed inside cell nuclei [Porter et al. 2007a, b]. Animal studies have shown that nanoparticles are more biologicallyactive due to their greater surface area per mass compared with larger-sized particles of the same chemistry [Oberdörster et al. 1992; 1994a,b; 2005a; Driscoll 1996; Lison et al. 1997; Brown et al. 2001; Duffin et al. 2002; Renwick et al. 2004; Barlow et al. 2005; Sager et al. 2008]. While this increased biological activity is a fundamental component to the utility of nanoparticles for industrial, commercial, and medical applications, the consequences of unintentional exposures of workers to nanoparticles are uncertain.

Research reported from laboratory animal studies and from epidemiological studies have lead to hypotheses regarding the potential adverse health effects of engineered nanomaterials. These hypotheses are based on the scientific literature of particle exposures in animals and humans. This literature has been recently reviewed [Donaldson et al. 2005; Maynard and Kuempel 2005; Oberdörster et al. 2005a, Donaldson et al. 2006; Kreyling et al. 2006]. In general, the particles used in past studies have not been characterized to the extent recommended for new studies in order to more fully understand the physicochemical properties of the particles that influence toxicity [Oberdörster et al. 2005b; Thomas et al. 2006]. As this research continues, more data will become available to support or refute the following hypotheses for engineered nanoparticles.

Hypothesis 1: Exposure to engineered nano- particles is likely to cause adverse health effects similar to ultrafine particles that have similar physical and chemical characteristics

Studies in rodents and humans support the hypothesis that exposure to ultrafine particles poses a greater respiratory hazard than exposure to the same mass of larger particles with a similar chemical composition. Studies ofexisting particles have shown adverse health effects in workers exposed to ultrafine particles (e.g., diesel exhaust particulate, welding fumes), and animal studies have shown that ultrafine particles are more inflammogenic and tumorigenic in the lungs of rats than an equal mass of larger particles of similar composition [Oberdörster and Yu 1990; Driscoll 1996; Tran et al. 1999, 2000]. If engineered nanoparticles have the same physicochemical characteristics that are associated with reported effects from ultrafine particles, they may pose the same health concerns.

Although the physicochemical characteristics of ultrafine particles and engineered nanoparticles can differ, the toxicologic and dosimetric principles derived from available studies may be relevant to postulating the health concerns for newly engineered particles. The biological mechanisms of particle-related lung diseases (i.e., oxidative stress, inflammation, and production of cytokines, chemokines, and cell growth factors) [Mossman and Churg 1998; Castranova 2000; Donaldson and Tran 2002] appear to be a consistent lung response for respirable particles including ultrafine or engineered nanoparticles [Donaldson et al. 1998; Donaldson and Stone 2003; Oberdörster et al. 2005a]. Toxicological studies have shown that the chemical and physical properties that influence the fate and toxicity of ultrafine particles may also be relevant to mechanisms influencing biological exposure and response to other nanoscale particles [Duffin et al. 2002; Kreyling et al. 2002; Oberdörster et al. 2002; Semmler et al. 2004; Nel et al. 2006].

Hypothesis 2: Surface area and activity and particle number may be better predictors ofpotential hazard than mass

The greater potential hazard may relate to thegreater number or surface area of nanoparticles compared with that for the same massconcentration of larger particles [Oberdörster et al. 1992, 1994a,b; Driscoll et al. 1996; Tran et al. 2000; Brown et al. 2001; Peters et al. 1997; Moshammer and Neuberger 2003; Sager et al. 2008]. This hypothesis is based primarily on the pulmonary effects observed in studies of rodents exposed to various types of ultrafine or fine particles (i.e., TiO_2, carbon black, barium sulfate, carbon black, diesel soot, coal fly ash, toner) and in humans exposed to aerosols, including nanoscale particles (e.g., diesel exhaust, welding fumes). These studies indicate that for a given mass of particles, relatively insoluble nanoparticles are more toxic than larger particles of similar chemical composition and surface properties. Studies of fine and ultrafine particles have shown that particles with less reactive surfaces are less toxic [Tran et al. 1999; Duffin et al. 2002]. However, even particles with low inherent toxicity (e.g., TiO_2) have been shown to cause pulmonary inflammation, tissue damage, and fibrosis at sufficiently high particle surface area doses [Oberdörster et al. 1992, 1994a,b; Tran et al. 1999, 2000].

Through engineering, the properties of nano- materials can be modified. For example, a recent study has shown that the cytotoxicity of water-soluble fullerenes can be reduced by several orders of magnitude by modifying the structure of the fullerene molecules (e.g., by hydroxylation) [Sayes et al. 2004]. These structural modifications were shown to reduce the cytotoxicity by reducing the generation of oxygen radicals—which is a probable mechanism by which cell membrane damage and death occurred in these cell cultures. Increasing the sidewall functionalization of SWCNT also rendered these nanomaterials less cytotoxic to cells in culture [Sayes et al. 2005]. Cytotoxicity studies with quantum dots haveshown that the type of surface coating can have a significant effect on cell motility and viability [Hoshino et al. 2004; Shiohara et al. 2004; Lovric et al. 2005]. Differences in the phase composition of nanocrystalline structures can influence their cytotoxicity; in a recent study comparing two types of TiO_2 nanoparticles exposed to UV radiation, anatase TiO_2 was more cytotoxic and produced more reactive species than did rutile TiO_2 with similar specific surface area (153 m^2g and 123 m^2g of TiO_2, respectively) [Sayes et al. 2006]. Reactive oxygen species were also associated with the cytotoxicity of TiO_2 nanoparticles to mouse microglia (brain cells) grown in culture [Long et al. 2006]. In contrast, in vitro generation of oxidant species is relatively low in purified SWCNT (contaminating metals removed), yet this material caused progressive interstitial fibrosis in vivo [Shvedova et al. 2004; 2005]. However, recent in vitro studies indicate that purified SWCNTs enhance proliferation and collagen production in fibroblasts [Wang et al. 2008]. Therefore, oxidant generation may not be the only mechanism driving the biological activity of nanomaterials.

The studies of ultrafine particles may provide useful data to develop preliminary hazard or risk assessments and to generate hypotheses for further testing. The studies in cell cultures provide information about the cytotoxic properties of nanomaterials that can guide further research and toxicity testing in whole organisms. More research is needed of the specific particle properties and other factors that influence the toxicity and disease development, including those characteristics that may be most predictive of the potentialsafety or toxicity of newly engineered nanomaterials.

6. Potential Safety Hazards

Very little is known about the safety risks that engineered nanomaterials might pose, beyond some data indicating that they possess certain properties associated with safety hazards in traditional materials. Based upon currently available information, the potential safety concerns most likely would involve catalytic effects or fire and explosion hazards if nanomaterials are found to behave similarly to traditional materials.

6.1. Fire and Explosion Risk

Although insufficient information exists to predict the fire and explosion risk associated with nanoscale powders, nanoscale combustible material could present a higher risk than a similar quantity of coarser material, given its unique properties [HSE 2004]. Decreasing the particle size of combustible materials can increase combustion potential and combustion rate, leading to the possibility of relatively inert materials becoming highly reactive in the nanometer size range. Dispersions of combustible nanomaterial in air may present a greater safety risk than dispersions of non- nanomaterials with similarcompositions. Some nanomaterials are designed to generate heat through the progression of reactions at the nanoscale. Such materials may present a fire hazard that is unique to engineered nanomaterials. In the case of some metals, explosion risk can increase significantly as particle size decreases.

The greater activity of nanoscale materials forms a basis for research into nanoenergetics. For instance, nanoscale Al/MoO_3 thermites ignite more than 300 times faster than corresponding micrometer-scale material [Granier and Pantoya 2004].

6.2. Risks of Catalytic Reactions

Nanoscale particles and nanostructured porous materials have been used as effectivecatalysts for increasing the rate of reactions or decreasing the necessary temperature for reactions to occur in liquids and gases. Depending on their composition and structure, some nanomaterials may initiate catalytic reactions that, based on their chemical composition, would not otherwise be anticipated [Pritchard 2004].

7. Exposure Assessment and Characterization

There are currently no national or international consensus standards on measurement techniques for nanomaterials in the workplace. If the qualitative assessment of a process has identified potential exposure points and leads to the decision to measure nano- materials, several factors must be kept in mind. Current research indicates that mass and bulk chemistry may be less important than particle size, surface area, and surface chemistry (or activity) for nanostructured materials [Oberdörster et al. 1992, 1994a,b; Duffin et al. 2002]. Research is ongoing into the relative importance of these different exposure metrics, and how to best characterize exposures to nanomaterials in the workplace. In addition, the unique shape and properties of some nanomaterials may pose additional challenges. For example, the techniques used to measure fiber concentrations in the workplace (e.g., phase contrast microscopy) would not be able to detect individual carbon nanotubes with diameters less than 100 nm nor bundles of carbon nanotubes with diameters less than 250 nm [Donaldson et al. 2006]. NIOSH and the National Institute of Standards and Technology (NIST) are collaborating on efforts to develop nanoscale reference materials for exposure assessment. Initial effort is focused on development of TiO_2 reference material.

7.1. Workplace Exposures

While research continues to address questions of nanomaterial toxicity, a number of exposure assessment approaches can be used to help determine worker exposures toairborne nanomaterials. These assessments can be performed using traditional industrial hygiene sampling methods including samplers placed at static locations (area sampling), samples collected in the breathing zone of the worker (personal sampling), or real-time devices or methods that can be personal or static. In general, personal sampling is preferred to ensure an accurate representation of the worker's exposure, whereas area samples (e.g., size-fractionated aerosol samples) and real-time (direct-reading) exposure measurements may be more useful for evaluating the need for improvement of engineering controls and work practices.

Many of the sampling techniques that are available for measuring nanoaerosols vary in complexity but can provide useful information for evaluating occupational exposures with respect to particle size, mass, surface area, number concentration, composition, and surface chemistry. Unfortunately, relatively few of these techniques are readily applicable to routine exposure monitoring. Research is ongoing into developing an analytical strategy for determining both TiO_2 surface area and titanium mass from 37-mm cassette filter samplers. Current measurement techniques are described below along with their applicability for monitoring nanometer aerosols.

For each measurement technique used, it is vital that the key parameters associated with the technique and sampling methodology be recorded when measuring exposure to nanoaerosols. This should include the response range of the instrumentation,whether personal or static measurements are made, and the location of all potential aerosol sources including background aerosols. Comprehensive documentation will facilitate comparison of exposure measurements using different instruments or different exposure metrics and will aid the re-interpretation of historic data as further information is developed on health- appropriate exposure metrics. Regardless of the metric and method selected for exposure monitoring, it is critical that measurements be taken before production or processing of a nanomaterial to obtain background nanoparticle exposure data. Measurements made during production or processing can then be evaluated to determine if there has been an increase in particle number concentrations in relation to background measurements and whether that change represents worker exposure to the nanomaterial. Table 7–1 gives a listing of instruments and measurement methods that can be used in the evaluation of engineered nanoparticle exposures.

Table 7–1. Summary of Instruments and Measurement Methods Used in the Evaluation of Nanomaterial Exposures*

Metric	Instrument or method	Remarks
Mass-Direct (total and/or elemental)	Size Selective Static Sampler	The only instruments offering a cut point around 100nm are cascade impactors (Berner-type low pressure impactors, or Micro orifice impactors). Allows gravimetric and chemical analysis of samples on stages below 100nm.
	TEOM® (Tapered Element Oscillat-ing Microbalance)	Sensitive real-time monitors such as the TEOM may be useable to measure nanoaerosol mass concentration on-line with a suitable size selective inlet.
	Filter collection and elemental analysis	Filters may be collected with size selective pre-samplers or open face. Elemental analysis (e.g., carbon, metals) for mass determination.
Mass-Indirect (calculation)	ELPI™ (Electrical Low Pressure Impactor)	Real time size-selective (aerodynamic diameter) detection of active surface area concentration giving aerosol size distribution. Mass concentration of aerosols can be calculated when particle charge and density are known or assumed.
	MOUDI (Micro-Orfice Uniform Deposit Impactor)	Real time size-selective (aerodynamic diameter) by cascade impaction.
	DMAS (Differential Mobility Analyzing System)	Real time size-selective (mobility diameter) detection of number concentration, giving aerosol size distribu-tion. Mass concentration of aerosols can be calculated when particle shape and density are known or assumed.
Number-Direct	CPC (Condensation Particle Counter)	CPCs provide real time number concentration measurements between their particle diameter detection limits. Without a nanoparticle pre-separator they are not specific to the nanometersize range. Some models have diffusion screen to limit top size to 1μm.
	OPC (Optical Particle Counter)	OPCs provide real time number concentration measurements between their particle diameter detection limits. Particle size diameters begin at 300nm and may go up to 10,000nm.
	DMAS and SMPS (Scanning Mobi-lity Particle Sizer)	Real time size-selective (mobility diameter) detection of number concentration giving number-based size distribution.
	Electron Microscopy	Off-line analysis of electron microscope samples can provide information on size-specific aerosol number concentration.
Number-Indirect	ELPI™ and MOUDI	Real time size-selective (aerodynamic diameter) detection of active surface-area concentration giving aerosol size distribution. Data may be interpreted in terms of number concentration. Size-selective samples may be further analyzed off-line.
Surface Area-Direct	Diffusion Charger	Real-time measurement of aerosol active surface-area. Active surface-area does not scale directly with geometric

Table 7–1. (Continued)

Metric	Instrument or method	Remarks
		surface-area above 100nm. Note that not all commercially available diffusion chargers have a response that scales with particle active surface-area below 100nm. Diffusion chargers are only specific to nanoparticles if usedwith appropriate inlet pre-separator.
	ELPI™ and MOUDI	Real-time size-selective (aerodynamic diameter) detection of active surface-area concentration. Active surface-area does not scale directly with geometric surface-area above 100nm.
	Electron Microscopy	Off-line analysis of electron microscope samples (previously collected on filters or other media) can provide information on particle surface-area with respect to size. TEM analysis provides directinformation on the projected area of collected particles which may be related to geometric area for some particles shapes.
Surface Area-Indirect (calculation)	DMAS and SMPS	Real time size-selective (mobility diameter) detection of number concentration. Data may be interpreted in terms of aerosol surface-area under certain circumsta-nces. For instance, the mobility diameter of open agglomerates has been shown to correlate with projected surface area.
	DMAS and ELPI™ used in parallel	Differences in measured aerodynamic and mobility can be used to infer particle fractal dimension which can be further used to estimate surface-area.

*Adapted from ISO/TR 12885

Note: Inherent to all air sampling instruments in this table is the fact that they cannot discriminate the nanoaerosol of interest from other airborne particles. Also, there is a general lack of validation regarding the response of these air sampling instruments to the full spectrum of nanoparticles that may be found in the workplace, including varieties of primary particles, agglomerates or aggregates, and other physical and chemical forms. A suite of nanoparticle reference materials are required to perform the needed validations.

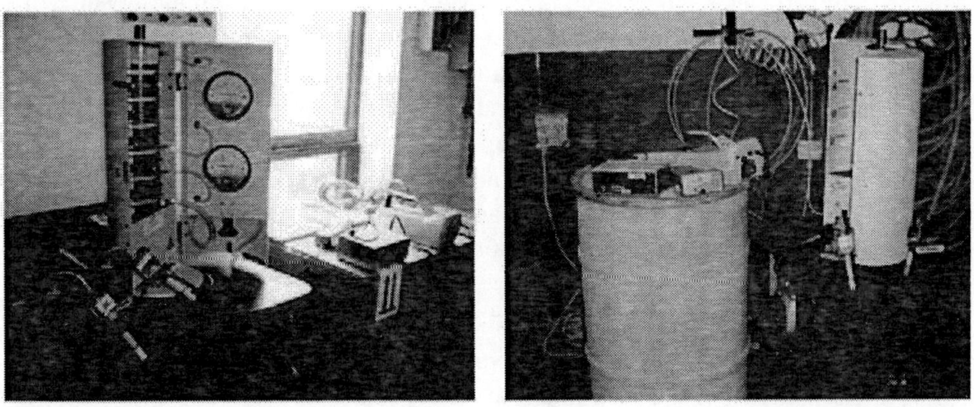

Figure 7–1. Examples of different sampling instruments used to measure occupational exposures to nanoparticles including the determination of real-time particle number concentrations and size-fractionated mass concentrations

7.1.1. Size-fractionated aerosol sampling

Studies indicate that particle size plays an important role in determining the potential adverse effects of nanomaterials in the respiratory system: by influencing the physical, chemical, and biological nature of the material; by affecting the surface-area dose of deposited particles; and by enabling deposited particles to more readily translocate to other parts of the body. Animal studies indicate that the toxicity of inhaled nanoparticles is more closely associated with the particle surface area and particle number than with the particle mass concentration when comparing aerosols with different particle size distributions. However, mass concentration measurements may be applicable for evaluating occupational exposure to nanometer aerosols where a good correlation between the surface area of the aerosol and mass concentration can be determined or if toxicity data based on mass dose are available for a specific nanoscale particle associated with a known process (e.g., diesel exhaust particulate).

Aerosol samples can be collected using inhalable, thoracic, or respirable samplers, depending on the region of the respiratory system most susceptible to the inhaled particles. Since prevailing information suggests that a large fraction of inhaled nanoparticles will deposit in the gas-exchange region of the lungs [ICRP 1994], respirable samplers would be appropriate. Respirable samplers will also collect a nominal amount of nanoscale particles that can deposit in the upper airways and ultimately be cleared or transported to other parts of the body.

Respirable samplers allow mass-based exposure measurements to be made using gravimetric and/or chemical analysis [NIOSH 1994]. However, they do not provide information on aerosol number, size, or surface-area concentration, unless the relationship between different exposure metrics for the aerosol (e.g., density, particle shape) has been previously characterized. Currently, no commercially available personal samplers are designed to measure the particle number, surface area, or mass concentration of nanoaerosols. However, several methods are available that can be usedto estimate surface area, number, or mass concentration for particles smaller than 100 nm.

The use of conventional impactor samplers to assess nanoparticle exposure is limited since the impaction collection efficiencies are 200–300 nm. Low-pressure cascade impactors that can measure particles to 50 nm and larger may be used for static sampling since their size and complexity preclude their use as personal samplers [Marple et al. 2001; Hinds 1999]. A personal cascade impactor is available with a lower aerosol cut point of250 nm [Misra et al. 2002], allowing an approximation of nanoscale particle mass concentration in the worker's breathing zone. For each method, the detection limits are on the order of a few micrograms of material on a filter or collection substrate [Vaughan et al. 1989]. Cascade impactor exposure data gathered from worksites where nanomaterials are being processed or handled can be used to make assessments as to the efficacy of exposure control measures.

7.1.2. Real-time aerosol sampling

The real-time (direct-reading) measurement of nanometer aerosol concentrations is limited by the sensitivity of the instrument to detect small particles. Many real-time aerosol mass monitors used in the workplace rely on light scattering from groups of particles (photometers). This methodology is generally insensitive to particles smaller than 100 nm [Hinds 1999]. Optical instruments that size individual particles and convert the measured distribution to a mass concentration are similarly limited to particles larger than 100 nm.

Quantitative information gained by optical particle counters may also be limited by relatively poor counting efficiencies at smaller particle diameters (i.e., less than 500 nm). These instruments are capable of operating within certain concentration ranges that, when exceeded, affect the count or mass determination efficiencies due to coincidence errors at the detector. Similarly, the response of optical particle counters may be material-dependent according to the refractive index of the particle. The Scanning Mobility Particle Sizer (SMPS) is widely used as a research tool for characterizing nanoscale aerosols although its applicability for use in the workplace may be limitedbecause of its size, cost, and the inclusion of a radioactive source. Additionally, the SMPS may take 2–3 minutes to scan an entire size distribution; thus, it may be of limited use in workplaces with highly variable aerosol size distributions, such as those close to a strong particle source. Fast (less than 1 second), mobility-based, particle-sizing instruments are now available commercially; however, having fewer channels, they lack the finer sizing resolution of the SMPS. The Electrical Low Pressure Impactor (ELPI) is an alternative instrument that combines diffusion charging and a cascade impactor with real-time (less than 1 second) aerosol charge measurements providing aerosol size distributions by aerodynamic diameter [Keskinen et al. 1992].

7.1.3. Surface-area measurements

Relatively few techniques exist to monitor exposures with respect to aerosol surface area. Particle surface is composed of internal surface area attributable to pores (cavities more deep than wide), external surface area due to roughness (cavities more wide than deep), and total surface area (the accessible area of all real particle surfaces). A standard gas adsorption technique (i.e., BET) is used to measure the total surface area of powders and can be adapted to measure the specific surface area (surface area per unit mass) of engineered nanomaterials [Brunauer et al. 1938]. However, surface-area analysis by gas adsorption requires relatively large quantities of material, is not element specific, and must be performedin a laboratory.

The first instrument designed to measure aerosol surface area was the epiphaniometer [Baltensperger et al. 1988]. This device measures the Fuchs, or active surface area, of the aerosols by measuring the attachment rate ofradioactive ions. For aerosols less than approximately 100 nm in size, measurement of the Fuchs surface area is probably a good indicator of external surface area (or geometric surface area). However, for aerosols greater than approximately 1 μm, the relationship with geometric particle surface area is lost [Fuchs 1964]. Measurements of active surface area are generally insensitive to particle porosity. The epiphaniometer is not well suited to widespread use in the workplace because of the inclusion of a radioactive source and the lack of effective temporal resolution.

This same measurement principle can be applied with the use of a portable aerosol diffusion charger. Studies have shown that these devices provide a good estimate of aerosol external surface area when airborne particles are smaller than 100 nm in diameter. For larger particles, diffusion chargers underestimate aerosol surface area. However, further research is needed to evaluate the degree of underestimation. Extensive field evaluations of commercial instruments are yet to be reported. However, laboratory evaluations with monodisperse silver particles have shown that two commercially available diffusion chargers can provide good measurement data on aerosol external surface area for particles smaller than 100 nm in diameter but underestimate the aerosol surface area for particles larger than 100 nm in diameter [Ku and Maynard 2005, 2006].

7.1.4. Particle number concentration measurement

Particle number concentration has been associated with adverse responses to air pollution in some human studies [Timonen et al. 2004; Ruckerl et al. 2005], while in toxicologic studies, particle surface area hasgenerally been shown to be a better predictor than either particle number, mass, or volume concentration alone [Oberdörster and Yu 1990; Tran et al. 1999; Duffin et al. 2002]. A two-variable dose metric of particle size and volume has been shown to be the best predictor of lung cancer in rats from various types of particles [Borm et al. 2004; Pott and Roller 2005]. This illustrates some of the complexity of interpreting existing data on particle dose metric and response. While adverse health effects appear to be more closely related with particle surface area, the number of particles depositing in the respiratory tract or other organ systems may also play an important role.

Aerosol particle number concentration can be measured relatively easily using Condensation Particle Counters (CPCs). These are available as hand-held static instruments, and they are generally sensitive to particles greater than 10–20 nm in diameter. Condensation Particle Counters designed for the workplace do not have discrete size-selective inputs, and so they are typically sensitive to particles less than 1 μm in diameter. Commercial size- selective inlets are not available to restrict CPCs to the nanoparticle size range; however, the technology exists to construct size- selective inlets based on particle mobility or possibly on inertial pre-separation. An alternative approach to estimating nanoparticle number concentrations using a CPC is to use the instrument in parallel with an optical particle counter (OPC). The difference in particle count between the instruments will provide an indication of particle number concentration between the lower CPC detectable particle diameter and the lowerOPC particle diameter (typically 300–500 nm).

A critical issue when characterizing exposure using particle number concentrationis selectivity.Nanoscale particles are ubiquitous in many workplaces, from sources such as combustion, vehicle emissions, and infiltration of outside air. Particle counters are generally insensitive to particle source or compositionmaking it difficult to differentiate between incidental and process-related nanoparticles using number concentration alone. In a study of aerosol exposures during a carbon black bagging process, Kuhlbusch et al. [2004] found that peaks in number concentration measurements were associated with emissions from fork lift trucks and gas burners in the vicinity, rather than with the process itself. In a similar manner, during an ultrafine particle mapping exercise in an automotive facility, Peters et al. [2006] found that direct gas- fired heating systems systematically produced high particle number concentrations throughout the facility when the heating system was in operation. Through follow up measurements, Heitbrink et al. [2007] found a high proportion of ultrafine particles produced from these burners, yet little if any mass was associated with their emissions. Other non-process ultrafine sources were identified in an adjacent foundry [Evans et al. 2008]. Together with roof mounted gas-fired heating units, additional sources included cigarette-smoking and the exhaust from a propane fueled sweeper vehicle, with the latter contributing a large fraction of the ultrafine particles. Although these issues are not unique to particle number concentration measurements, orders of magnitude difference can exist in particle number concentrations depending on concomitant sources of particle emissions.

Although using nanoparticle number concentration as an exposure measurement may not be consistent with exposure metrics being used in animal toxicity studies, suchmeasurements may be useful for identifying nanoscale particle emissions and determining the efficacy of control measures. Portable CPCs are capable of measuring localized particle concentrations allowing the assessment of particle releases occurring at various processes and job tasks [Brouwer et al. 2004].

7.1.5. Surface-area estimation

Information about the relationship between different measurement metrics can be used for approximating particle surface area. If the size distribution of an aerosol remains consistent, the relationship between particle number, surface area, and mass metrics will be constant. In particular, mass concentration measurements can be used for deriving surface-area concentrations, assuming the constant of proportionality is known. This constant is the specific surface area (surface to mass ratio).

Size distribution measurements may be obtained through the collection of filter samples and analysis by transmission electron microscopy to estimate particle surface area. If the measurements are weighted by particle number, information about particle geometry will be needed to estimate the surface area of particles with a given diameter. If the measurements are weighted by mass, additional information about particle density will be required. Estimates of particle-specific surface area from geometric relation with external particle dimensions depends upon the morphology regime of the material of interest and is only appropriate for smooth, regularly shaped, compact particles [Stefaniak et al. 2003; Weibel et al. 2005]. For example, Weibel et al. [2005] report that estimates of ultrafine TiO_2 surface areadetermined using a geometric relationship with the physical particle size(using TEM) were 50% lower than measured using nitrogen gas adsorption.

If the airborne aerosol has a lognormal size distribution, particle surface-area concentration can be derived using three independent measurements. An approach has been proposed using three simultaneous measurements of the aerosol that included mass concentration, number concentration, and charge [Woo et al. 2001]. With knowledge of the response function of each instrument, minimization techniques can be used to estimate the parameters of the lognormal distribution leading to the three measurements used in estimating the particle surface area.

An alternative approach has been proposed whereby independent measurements of particle number and mass concentration are made, and the surface area is estimated by assuming the geometric standard deviation of the (assumed) lognormal distribution [Maynard 2003]. This method has the advantage of simplicity by relying on portable instruments that can be used in the workplace. Theoretical calculations have shown that estimates may be up to a factor of 10 different from the actual particle surface area, particularly when the aerosol has a bimodal distribution. Field measurements indicate that estimates are within a factor of 3 of the active surface area, particularly at higher concentrations. In workplace environments, particle surface- area concentrations can be expected to span up to 5 orders of magnitude; thus, surface- area estimates may be suited for initial or preliminary appraisals of occupational exposure concentrations.

Although such estimation methods are unlikely to become a long-term alternative to more accurate methods, they may provide a viable interim approach to estimating the surface area of nanoscale particles in theabsence of precise measurement data. Additional research is needed on comparing methods used for estimating particle surface area with a more accurate particle surfacearea-measurement method. NIOSH is conducting research in this area and will communicate results as they become available.

7.1.6. Particle numberconcentration mapping

To better understand particle sources and contaminant migration, some investigators have adopted an aerosol mapping technique, which integrated measurements of respirable mass, ultrafine particle number, and active surface-area concentrations in automotive manufacturing facilities [Peters et al. 2006; Heitbrink et al. 2007, 2008; Evans et al. 2008]. The process relies on portable aerosol sampling instrumentation for simultaneous measurements at predetermined positions throughout a facility. The technique is somewhat measurement-intensive but is useful for locating contaminant sources and determining the extent of contaminant migration. Leaks and other less obvious particle sources have been identified in this way and the procedure provides a powerful tool for facility staff to target their contaminant control approaches most effectively. This technique relies on successive measurements at various locations, making facilities with continuous processes or those likelyto achieve steady state particle number concentrations most appropriate for this approach. The approach is less successful for facilities with batch processes or those likely to experience rapid concentration changes as, depending on where in the measurement cycle the release occurs, it may be overlooked. A high degree of variability between mapping events is expected infacilities where sporadic or batch processing occurs.

7.2. Sampling Strategy

Currently, there is not one sampling method that can be used to characterize exposure to nanoscale aerosols. Therefore, any attempt to characterize workplace exposure to nanomaterials must involve a multifaceted approach incorporating many of the sampling techniques mentioned above. Brouwer et al. [2004] recommend that all relevant characteristics of nanomaterial exposure be measured, and a sampling strategy similar to theirs would provide a reasonable approach to characterizing workplace exposure. NIOSH has developed the Nanoparticle Emission Assessment Technique (NEAT) to qualitatively determine the release of engineered nanomaterials in the workplace (see Appendix). This approach may be helpful to others for the initial evaluation of workplaces where engineered nanomaterials are manufactured or used. If material release is found and if resources allow, then a more comprehensive and quantitative approach may be adopted [Methner et al. 2007].

The first step to characterizing workplace exposures would involve identifying the source of nanomaterial emissions. A CPC used inparallel with an OPC provides acceptable capability for this purpose. It is critical to determine ambient or background particle counts before measuring particle counts during the manufacturing, processing, or handling of engineered nanomaterials. However, investigators need to be aware that background nanoscale particle counts can vary both spatially and temporally depending on the unique conditions of the workplace. Subtraction of background nanoscale particle counts will be most challenging in these situations. In cases where nanomaterial handling or processing operations contribute only small elevations in particle counts, it may not be possible to adequately characterize these increases, particularly if the background particle count is relatively high.

If nanomaterials are detected in the process area at elevated concentrations relative to background particle number concentrations, then a pair of filter-based, area air samples should be collected for particle analysis via transmission electron microscopy (TEM) and for determining mass concentration. Transmission electron microscopycan provide an estimate of the particle size distribution and, if equipped with an energy dispersive X-ray analyzer (EDS), a determination of elemental composition can be made to identify the nanomaterial (see Figure 7–2).

Analysis of filters for mass determination of air contaminants of interest can help identify the source of the particles. Standard analytical chemical methodologies (e.g., NMAM 5040 for carbon, NMAM 7303 for metals) should be employed [NIOSH 1994].

The combination of particle counters and samples for chemical analysis allows for an assessment of worker exposure to nanomaterials (see Figure 7–3) and the characterization of the important aerosol metrics. However, since this approach relies primarily on static or area sampling, some uncertainty will exist in estimating worker exposures.

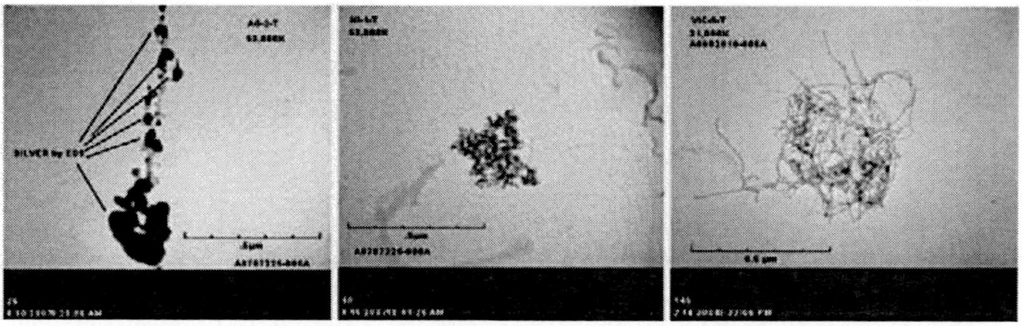

Figure 7–2. Photomicrographs of airborne engineered nanomaterials (airborne samples of engineered nanoparticles of silver, nickel, and MWCNT analyzed by TEM and EDS)

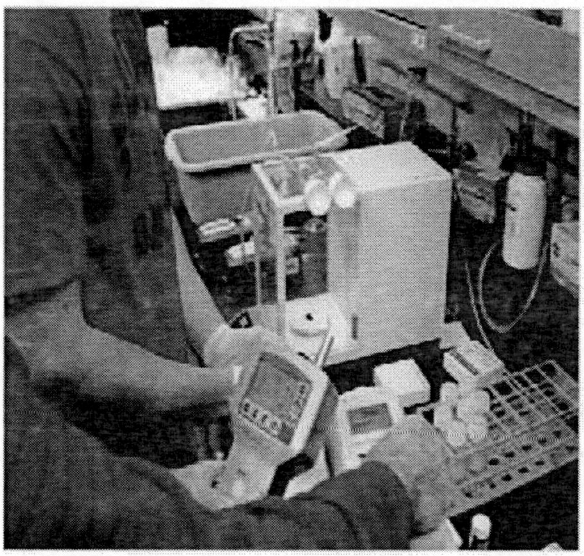

Figure 7–3. Combined use of the OPC, CPC, and two filter samples to determine the presence of nanomaterials

8. Guidelines for Working with Engineered Nanomaterials

Engineered nanomaterials are diverse in their physical, chemical, and biological nature. The processes used in research, material development, production, and use or introduction of nanomaterials have the potential to vary greatly. Until further information on the possible health risks and extent of occupational exposure to nano- materials becomes available, interim protective measures should be developed and implemented. These measures should focus on the development of engineering controls and safe working practices tailored to the specific processes and materials where workers might be exposed. Hazard information that is available about common materials being manufactured in the nanoscale range (e.g., TiO_2, beryllium) should be considered as a starting point in developing appropriate controls and work practices.

The following recommendations are designed to aid in the assessment and control of workplace exposures to engineered nano- materials. Using a hazard-based approach to evaluate exposures and for developing precautionary measures is consistent with good occupational safety and health practices [The Royal Society and The Royal Academy of Engineering 2004; Schulte et al. 2008].

8.1. Potential for Occupational Exposure

Few workplace measurement data exist on airborne exposure to nanomaterials that are purposely produced and not incidental to an industrial process. In general, it is likely that processes generating nanomaterials in thegas phase (after removal of the nanomaterial from an enclosed generation system), or using or producing nanomaterials as powders or slurries/suspensions/solutions (i.e., in liquid media), pose the greatest risk for releasing nanoparticles. In addition, maintenance on production systems (including cleaning and disposing of materials from dust collection systems) is likely to result in exposure to nanoparticles if deposited nanomaterials are disturbed. Exposures associated with waste streams containing nanomaterials may also occur.

The magnitude of exposure to nanomaterials when working with nanopowders depends on the likelihood of particles being released from the powders during handling. NIOSH is actively conducting research to quantitatively determine how various nanomaterials are dispersed in the workplace. Studies on exposure to SWCNTs and MWCNTs have indicated that the raw material may release visible particles into the air when handled, that the particle size of the agglomerate can be a few millimeters in diameter, and that the release rate of inhalable and respirable particles is relatively low (on a mass or number basis) compared with other nanopowders. Maynard et al. [2004] reported concentrations of respirable dust from 0.007 to 0.053 mg/m^3 when energy was applied (vortexing) to bulk SWCNT forapproximately 30 minutes. Similar findings were reported by Han et al. [2008] at a laboratory producing MWCNTs in which exposure concentrations as high as 0.4 mg/m^3 were observed prior to the implementation of exposure controls. In a health hazard evaluation conducted by NIOSH at auniversity-based research laboratory the potential release of airborne carbon nanotubes (CNFs) was observed at various processes [Methner et al. 2007]. General area exposure measurements indicated slight increases in airborne particle number and mass concentrations relative to background measurements during the transfer of CNFs prior to weighing and mixing, and during wet saw cutting of a composite material. Since data are lacking on the generation of inhalable/ respirable particles during the production and use of engineered nanomaterials, further research is required to determine exposures under various conditions. NIOSH researchers are conducting both laboratory and field-based evaluations in order to address some of these knowledge gaps.

Devices comprised of nanostructures, such as integrated circuits, pose a minimal risk of exposure to nanomaterials during handling. However, some of the processes used in their production may lead to exposure to nanomaterials (e.g., exposure to commercial polishing compounds that contain nanoscale particles, exposure to nanoscale particles that are inadvertently dispersed or created during the manufacturing and handling processes). Likewise, large-scale components formed from nanocomposites will most likely not present significant exposure potential. However, if such materials are used or handled in such a manner that can generate nanoparticles (e.g., cutting, grinding) or undergo degradation processes that lead to the release of nanostructured material, then exposure may occur by the inhalation, ingestion, and/or dermal penetration of these particles.

8.2. Factors Affecting Exposure to Nanomaterials

Factors affecting exposure to engineered nanomaterials include the amount of material being used and whether the material can be easily dispersed (in the case of a powder) or form airborne sprays or droplets (in the case of suspensions). The degree of containment and duration of use will also influence exposure. In the case of airborne material, particle or droplet size will determine whether the material can enter the respiratory tract and where it is most likely to deposit. Respirable particles are those capable of depositing in the alveolar (gas exchange) region of the lungs, which includes particles smaller than approximately 10 μm in diameter [Lippmann 1977; ICRP 1994; ISO 1995]. The proportion of inhaled nanoparticles likely to deposit in any region of the human respiratory tract is approximately 30%–90% depending on factors such as breathing rate and particle size. Up to 50% of nanoparticles in the 10–100 nm size range may deposit in the alveolar region, while nanoparticles smaller than 10 nm are more likely to deposit in the head and thoracic regions [ICRP 1994]. The mass deposition fraction of inhaled nanoparticles in the gas-exchange region of the lungs is greater than that for larger respirable particles.

At present there is insufficient information to predict all of the situations and workplace scenarios that are likely to lead to exposure to nanomaterials. However, there are some workplace factors that can increase the potential for exposure:

- working with nanomaterials in liquid media without adequate protection (e.g., gloves)
- working with nanomaterials in liquid during pouring or mixing operationsor where a high degree of agitation is involved
- generating nanomaterials in the gas phase in non-enclosed systems
- handling (e.g., weighing, blending, spraying) powders of nanostructured materials
- maintenance on equipment and processes used to produce or fabricate nanomaterials

- cleaning up spills or waste material
- cleaning dust collection systems used to capture nanoparticles
- machining, sanding, drilling of nano- materials, or other mechanical disruptions of nanomaterials can potentially lead to the aerosolization of nanoparticles.

8.3. Elements of a Risk Management Program

Given the limited information about the health risks associated with occupational exposure to engineered nanomaterials, appropriate steps should be taken to minimize the risk of worker exposure through the implementation of a risk management program [Schulte et al. 2008]. Risk management programs for nanomaterials should be seen as an integral part of an overall occupational safety and health program for any company or workplace producing or using nanomaterials or nanoenabled products. A critical element of the program should be the capability to anticipate new and emerging risks (hazard determination) and whether they are linked to changes in the manufacturing process, equipment, or the introduction of new materials. This will require an ongoing assessment of the potential risks to workers (riskevaluation) through the systematic collection of job and product information so that determinations can be made regarding scenarios (e.g., laboratory research, production and manufacture, nanoenabled product use) that place the worker in contact with nanomaterials (see Figure 8–1). This assessment should be an ongoing cyclic process that provides feedback on potential sources of exposure and solutions taken to correct those problems. For example, operations and job tasks that have the potential to aerosolize nanomaterials (e.g., handling dry powders, spray applications) deserve more attention and more stringent controls than those where the nano- materials are imbedded in solid or liquid matrices. Elements of the risk management program should include guidelines for installing and evaluating engineering controls (e.g., exhaust ventilation, dust collection systems), the education and training of workers in the proper handling of nanomaterials (e.g., good work practices), and the selection and use of personal protective equipment (e.g., clothing, gloves, respirators).

When controlling potential exposures within a workplace, NIOSH has recommended a hierarchical approach to reduce worker exposures (see Table 8–1) [NIOSH 1990]. The philosophical basis for the hierarchy of controls is to eliminate the hazard when possible (i.e., substitute with a less hazardous material) or, if not feasible, control the hazard at or as close to the source as possible.

8.3.1. Engineering Controls

If the potential hazard can not be eliminated or substituted with a less hazardous or non-hazardous substance, then engineering controls should be installed and tailored to the process or job task. The type of engineering control used should take into account information on the potential hazardous properties of the precursor materials and intermediates as well as those of the resulting nanomaterial. In light of current scientific knowledge about the generation, transport, and capture of aerosols [Seinfeld and Pandis 1998; Hinds 1999], airborne exposure to nanomaterials can most likely be controlled at most processes and job tasks using a wide variety of engineering control techniques similar to those used in reducing exposures to general aerosols [Ratherman 1996; Burton 1997].

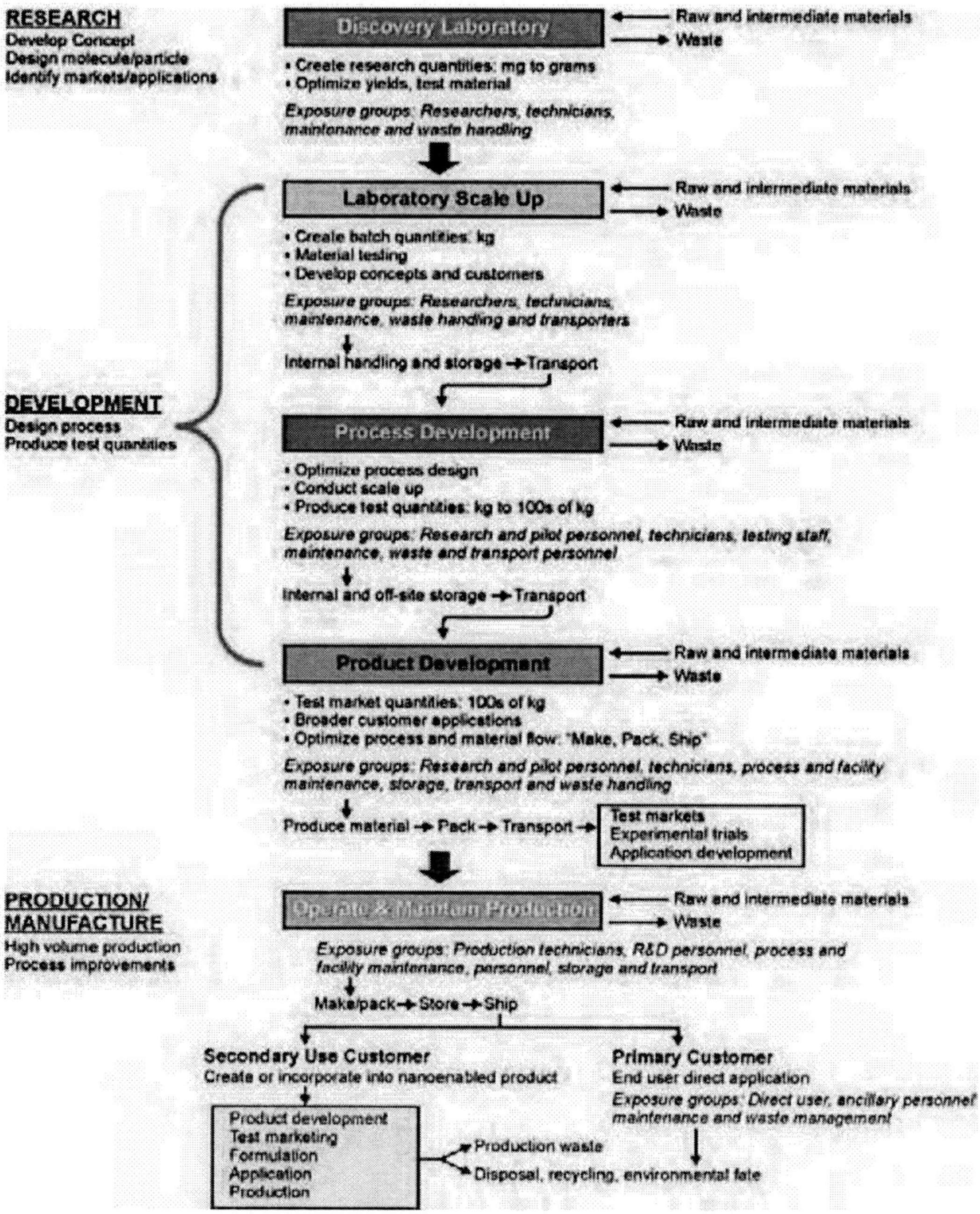

Figure 8–1. Workplaces with potential for occupational exposure to engineered nanomaterials.The figure illustrates the life cycle of nanomaterials from laboratory research development through product development, use, and disposal. Each step of the life cycle represents opportunities for potential worker exposure to nanomaterials. Adapted from Schulte et al. 2008a.

Engineering control techniques such as source enclosure (i.e., isolating the generation source from the worker) and local exhaust ventilation systems should be effective for capturing airborne nanomaterials, based on what is known of nanomaterial motion and behavior in air (see Figure 8–2). The quantity of the bulk nanomaterial that is synthesized or handled in the manufacture of a product will significantly influence the selection of the exposure controls.

Other factors that influence selection of engineering controls include the physical form of the nanomaterial and task duration and frequency. For instance, working with nanomaterial in the slurry form in low quantities would require a less rigorous control system than those that would be required for large quantities of nanomaterials in a free or fine powder form (see Figure 8–3). Unless cutting or grinding occurs, nanomaterials that are not in free form (encapsulated in a solid, nanocomposites, and surface coated materials) typically wouldn't require engineering controls.

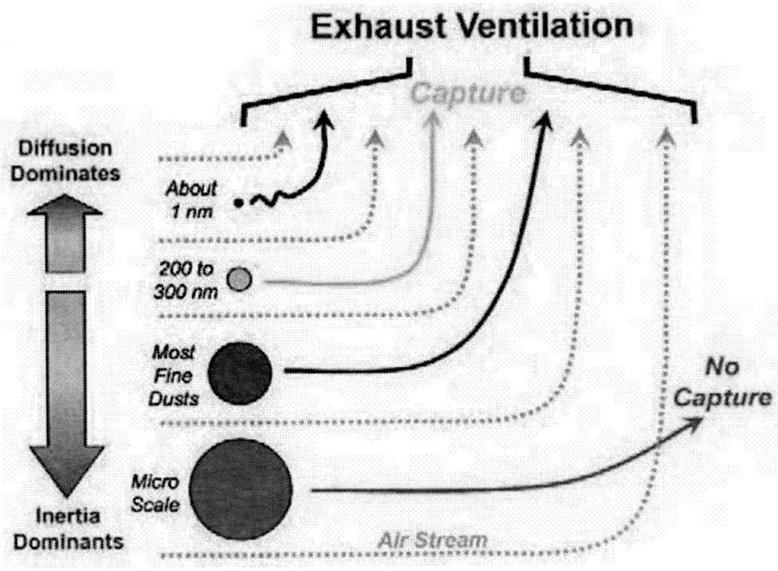

Figure 8–2. Exposure control of particles (illustration of how particle diameter-related diffusion and inertia influence particle capture efficiency in a ventilation system). Particles with a diameter of 200–300 nm have minimal diffusion and inertial properties and are easily transported by moving air and captured. Particle motion by diffusion increasingly dominates as particle diameter decreases below 200 nm. The inertial behavior of larger particles, especially those ejected from energetic processes such as grinding, increases significantly with particle diameter, enabling them to cross the streamlines of moving air and avoid capture. Adapted from Schulte et al. 2008a.

Table 8–1. Hierarchy of Exposure Controls*

Control method	Process, equipment, or job task
1. Elimination	Change design to eliminate hazard
2. Substitution	Replace a high hazard for a low hazard
3. Engineering	Isolation/enclosure, ventilation (local, general)
4. Administrative	Procedures, policies, shift design
5. Personal protective equipment	Respirators, clothing, gloves, goggles, ear plugs

*Control methods are typically implemented in this order to limit worker exposures to an acceptable concentration (e.g., occupational exposure limit or other pre-established limit).

Sources: Plog et al. 2002; NIOSH 1990.

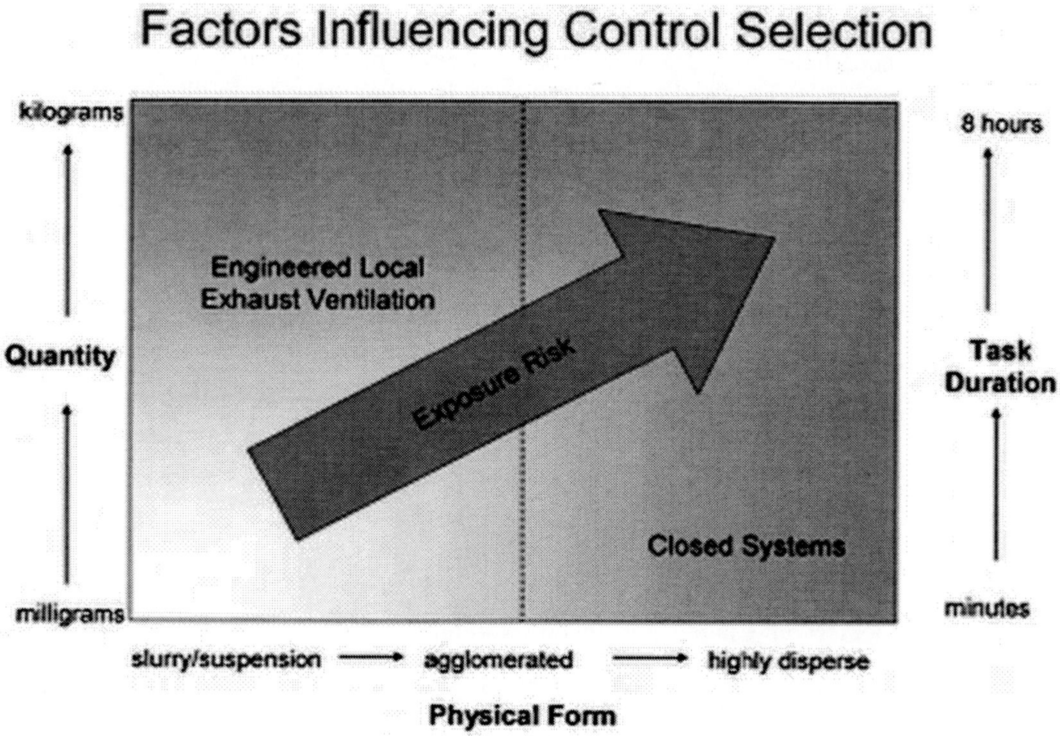

Figure 8–3. Factors influencing control selection. Several factors influence the selection of exposure controls for nanomaterials including quantity of nanomaterial handled or produced, physical form, and task duration. As each one of theses variables increase, exposure risk becomes greater as does the need for more efficient exposure control measures.

Handling research quantities typically occurs in laboratories with ventilation controls. Since quantities are small, local containment and control can be applied, such as low-flow vented work stations and small glove box chambers. However, as quantities are increased, care must be taken to reduce the amount of nanomaterial that is released from the process equipment and to prevent the migration of nanomaterials into adjacent rooms or areas. For example, the installation of local exhaust ventilation at a reactor used to make nanoscale engineered metal oxides and metals was found to reduce nanoparticle exposures by 96% (mean particle number concentration) [Methner 2008]. The use of exhaust ventilation systems should be designed, tested, and maintained using approaches recommended by the American Conference of Governmental Industrial Hygienists [ACGIH 2001].

A secondary but nonetheless important issue concerning the control of nanoparticle emissions is that of product loss. The properties of nanomaterials, along with theunique methods that may be employed for producing them, may mean that traditional exhaust ventilation may be more energetic than necessary for removing incidentally released nanoscale particles. For this reason, engineering controls need to be applied judiciously to ensure protection of workers without compromising production.

8.3.2. Dust collection efficiency of filters

Current knowledge indicates that a well-designed exhaust ventilation system with a HEPA filter should effectively remove nano- particles [Hinds 1999]. Limited studies have reported the effi cacy of filter media typically found in control systems (including respirators) in capturing nanoparticles. The dearth of data on filtration performance against nanoparticles (in particular nanoparticles smaller than 20 nm) is primarily due to the challenges in generating and quantifying particles in those size ranges. Despite these limitations, the results of some studies [Van Osdell et al. 1990] using different filter media challenged with monodispersed aerosols (silver 4–10 nm and dioctylphthalate 32–420 nm) were in agreement with classical single-fiber theory showing an increase in filtration efficacy for smaller size particles. No evidence for particle thermal rebound was observed. Similar results have been recently reported by Kim et al. [2007] using different filter media challenged with particles ranging in size from 2.5–20 nm, indicating that other filter medias—including those used in air purifying respirators—would behave similarly.

If HEPA filters are used in the dust collection system, they should be coupled with well-designed filter housings. If the filter is improperly seated within the housing, nanoparticles have the potential to bypass the filter, leading to fi lter effi ciencies much less than predicted [NIOSH 2003].

8.3.3. Work practices

An integral step in establishing good workpractices is having knowledge of the potentialhazards in the workplace and developing formal procedures that describe actions to be taken to ensure the protection of workers. Incorporated in these procedures should be guidelines for good work practices intended to minimize worker exposure to nanomaterials and other potentially hazardous chemicals. Management should systemically review and update these procedures. Actions taken to resolve and/or improve workplace conditions should be routinely conveyed by management to workers.

Good practices for management

- Educating workers on the safe handling of engineered nano-objects or nano-object-containing materials to minimize the likelihood of inhalation exposure and skin contact.
- Providing information, as needed, on the hazardous properties of the precursor materials and those of the resulting nanomaterials product with instruction on measures to prevent exposure.
- Encouraging workers to use hand- washing facilities before eating, smoking, or leaving the worksite.
- Providing additional control measures (e.g., use of a buffer area, decontamination facilities for workers if warranted by the hazard) to ensure that engineered nanomaterials are not transported outside the work area [US DOE 2007].
- Providing facilities for showering and changing clothes to prevent the inadvertent contamination of other areas (including take-home) caused by the transfer of nanomaterials on clothing and skin.

Good practices for workers

- Avoiding handling nanomaterials inthe open air in a *'free particle"* state.

- Storing dispersible nanomaterials, whether suspended in liquids or in a dry particle form in closed (tightly sealed) containers whenever possible.
- Cleaning work areas at the end of each work shift, at a minimum, using either a HEPA-filtered vacuum cleaner or wet wiping methods. Dry sweeping or air hoses should not be used to clean work areas. Cleanup should be conducted in a manner that prevents worker contact with wastes. Disposal of all waste material should comply with all applicable Federal, State, and local regulations.
- Avoiding storing and consuming food or beverages in workplaces where nanomaterials are handled.

8.3.4. Personal protective clothing

Currently, there are no generally acceptable guidelines available based on scientific data for the selection of protective clothing or other apparel against exposure to nanomaterials.This is due in part to minimal data being available on the efficacy of existing protective clothing, including gloves. In any case, although nanoparticles may penetrate the epidermis, there has been little evidence to suggest that penetration leads to disease; and no dermal exposure standards have been proposed. However, based on a recent survey of nanotechnology workplaces [ICON 2006], 84% of employers recommended personal protective equipment and clothing for employees working with nanomaterials. These recommendations were generally based on conventional occupational hygiene practicesbut also varied with the size of the company, the type of nanomaterials being handled, and the commercial sector. While some guidelines on the use of protective clothing and gloves have been developed by organizations for use in their own laboratories (e.g., US DOE 2007) or countries (e.g., British Standards Institute BSI 2008) or by consensus standards development organizations (e.g., ASTM, 2007), these are generally based upon good industrial hygiene practices rather than scientific data specific to nanomaterials.

A challenge to making appropriate recommendations for dermal protection against nanoparticles is the need to strike a balance between comfort and protection. Garments that provide the highest level of protection (e.g., an impermeable Level A suit) are also the least comfortable to wear for long periods of time, while garments that are probably the least protective (e.g., thin cotton lab coat) are the most breathable and comfortable for employees to wear. The two primary routes of exposure to particulates for workers using protective clothing are direct penetration through the materials and leakage through gaps, seams, defects, and interface and closure areas [Schneider et al. 1999, 2000]. The relative contributions from these two inward leakage sources are not well-understood. NIOSH has an active research program designed to assess the efficacy of barrier materials and ensembles for protection against particulate hazards, including nanoparticles.

The lack of available data is further complicated by the limitations and difficulties of current test methods, which fall into two basic categories: penetration tests on material swatches to determine barrier efficiency and system-level aerosol testing to determine product ensemble integrity. The former are usually bench-scale testing methods, whilethe latter require an exposure chamber that is large enough for at least one human test subject or mannequin. Chamber design requirements for system level aerosol testing have been reviewed by Gao et al. [2007]. Little scientific data exists, but some systems level test methods are available. ISO standard method 13982 [ISO 2004a] and EN standard method 943

[CEN 2002] specify the use of sodium chloride (NaCl) with a mass median aerodynamic diameter (MMAD) of 0.6 μm to determine the barrier efficiency of protective clothing against aerosols of dry, fine dusts. The standard method issued by National Fire Protection Association [2007] is a method that is not dependent on filtration-based approaches. Penetration of fluorophore-impregnated silica particles with a MMAD of 2.5 μm and a geometric standard deviation of 2.6 are qualitatively visualized by black light that causes the fluorescent glow of the challenge aerosol particles. Note that the polydisperse particle challenges used in these methods include a large number of nanoscale particles when measured by count rather than by mass.

Particle penetration test methods can be further categorized into those that are analogous to the process used in respirator filter testing and those that are not dependent on filtration-based approaches. Test methods that involve measuring aerosol concentrations using a sampling flow rate do not mimic in-situ situations because the skin does not "breathe." Standardized methodology is needed that is not dependent on filtration-based approaches for examining the overall barrier-effectiveness of the full protective clothing ensemble for different materials to particulate hazards. In this respect, NIOSH has presented preliminary results [Wang and Gao2007] on development of a magnetic passive aerosol sampler for more accurate determination of particle penetrationthrough protective clothing. NIOSH is conducting research in this area and will communicate results as they become available.

The bulk of the penetration data available on clothing has been done with filtration based testing. One study found that penetration levels of 30–2,000-nm-sized potassium chloride particles through an unidentified military garment ranged from about 20%–60%, with the maximum penetration occurring in the range of 100–400 nm [Hofacre 2006]. Another group of researchers studied the barrier efficiency of 10 unidentified fabric samples (woven, nonwoven, and laminated fabrics) using 477-nm-sized latex spheres at a flow rate of 1.8 cm/second [Shavlev et al. 2000]. Particle penetration measurements ranged from 0%–54%, with three of the fabrics exhibiting a measurable pressure drop and having penetration levels less than 1%. In general, these findings suggest that increased external air pressure (e.g., from wind) results in increased particle penetrations. Thus, only impermeable barrier materials are likely to provide complete barrier protection against aerosol penetration. Body movement (i.e., bellows effect) can also impact penetration [Bergman et al. 1989]. NIOSH will theoretically and empirically investigate wind-driven nanoparticle penetration through protective clothing in an attempt to obtain a predictive model based upon single-fiber theory. Results will be communicated as they become available.

Another widely used test method incorporates testing with nanoscale particles in solution, and therefore also provides some indication of the effectiveness of protective clothing to nanoparticles. ASTM standard F1671–03 [ASTM 2003] and ISO standard 16604 [ISO 2004b] specify the use of a 27-nm bacteriophage to evaluate the resistance ofmaterials used in protective clothing from the penetration of blood-borne pathogens. One study [Edlich et al. 1999] evaluated the integrity of powder-free examination gloves and found that no bacteriophage penetration was detected for powder-free nitrile gloves, powder-free latex gloves, nor polyvinyl chloride synthetic gloves.

Based upon the uncertainty of the health effects of dermal exposure to nanoparticles, it is prudent to consider using protective equipment, clothing, and gloves to minimize dermal exposure, with particular attention given to preventing exposure of nanomaterials to abraded or lacerated skin. Until scientific data exist specific to the performance of protective clothing and gloves against nanomaterials, current industrial hygiene best practices should be followed.

8.3.5. Respirators

The use of respirators is often required when engineering and administrative controls do not adequately keep worker exposures to an airborne contaminant below a regulatory limit or an internal control target. Currently, there are no specific exposure limits in the United States for airborne exposures to engineered nanomaterials although occupational exposure limits and guidelines exist for airborne particles of similar chemical composition regardless of particle size. Current scientific evidence indicates that nanoparticles may be more biologically re-active than larger particles of similar chemical composition and thus may pose a greater health risk when inhaled. In determining the need for respirators, it would therefore be prudent to consider current exposure limits or guidelines (e.g., OSHA PELs,NIOSH RELs, ACGIH TLVs) for larger particles of similar composition, existing toxicologicdata on the specific nanoparticle, and the likelihood of worker exposure (e.g., airborne concentration, time exposed, job task).

The decision to institute respiratory protection should be based on a combination of professional judgment and the results of the hazard assessment and risk management practices recommended in this document. The effectiveness of administrative, work-practice, and engineering controls can be evaluated using the measurement techniques described in Chapter 7 Exposure Assessments and Characterization. If worker exposure to airborne nanomaterials remains a concern after instituting control measures, the use of respirators can provide further worker protection. Several classes of respirators exist that can provide different levels of protection when properly fit tested on the worker. Table 8–2 lists various types of particulate respirators that can be used; information is also provided on the level of exposure reduction that can be expected along with the advantages and disadvantages of each respirator type. To assist respirator users, NIOSH has published the document *NIOSH Respirator Selection Logic (RSL)* that provides a process that respirator program administrators can use to select appropriate respirators [NIOSH 2004] (see www.cdc. gov/niosh/docs/2005-100/default.html). As new toxicity data for individual nanomaterials become available, NIOSH will review the data and make recommendations for respirator protection.

When respirators are required for use in the workplace, the Occupational Safety and Health Administration (OSHA) respiratory protection standard [29 CFR 1910.134] requires that a respiratory program be established that includes the following program elements: (*1*) an evaluation of the worker'sability to perform the work while wearing a respirator, (*2*) regular training of personnel, (*3*) periodic environmental monitoring, (*4*) respirator fit testing, and (*5*) respirator maintenance, inspection, cleaning, and storage. The standard also requires that the selection of respirators be made by a person knowledgeable about the workplace and the limitations associated with each type of respirator. OSHA has also issued guidelines for employers who choose to establish the voluntary use of respirators [29 CFR 19 10.134 Appendix D].

Table 8–2 lists the NIOSH assigned protection factors (APF) for various classes of respirators. The APF is defined as the minimum anticipated protection provided by a properly functioning respirator or class of respirators to a given percentage of properly fitted and trained users. The APF values developed by NIOSH are based in part on laboratory studies and take into consideration a variety of factors including the inward leakage caused by penetration through the filter and leakage around the respirator face seal. The relative contributions of these two sources of inward leakage are critical because for many applications the predominant source of exposure to the respirator wearer results from leakage around the face seal (due to a poor fit) and not penetration directly through the filter media. In 2006, OSHA published updated APF values that supersede the NIOSH APF values [Federal Register 2006]. In general there is good agreement between the NIOSH and OSHA APF values, but management should consult the OSHA standard prior to using the values in Table 8–2 directly.

NIOSH is not aware of any data specificto respirator face seal leakage of nanoparticles. However, numerous studies havebeen conducted on larger particles and on gases/vapors with one total inward leakage (TIL) study that used nanoparticles. For example, work done by researchers at the U.S. Army RDECOM on a head-form showed that mask leakage (i.e., simulated respirator fit factor) measured using submicron aerosol challenges (0.72 ѕm polystyrene latex spheres) was representative of vapor challenges such as sulfur hexafluoride (SF_6) and isoamyl acetate (IAA) [Gardner et al. 2004]. Other studies using particles larger than 100 nm have shown that face seal leakage can be affected by particle size, however, the impact of this is still the subject of some debate. A recently completed laboratory study to measure TIL protection factors of four NIOSH certified N95 filtering facepiece respirator models donned by human test subjects exposed to 40–1,300 nm particles found that the minimal protection factors were observed for particles between 80–200 nm [Lee 2008]. The geometric mean of the protection factors for all four models across all particle sizes tested was 21.5; but wide model-to-model variation was observed. NIOSH is conducting a laboratory study to determine whether nanoparticle face seal leakage is consistent with the leakage observed for larger particles and gases/ vapors. Results will be communicated as they become available.

NIOSH certifies respirators in accordance with 42 CFR Part 84. As noted earlier, the NIOSH RSL contains a process for selecting respirators for protection against particular hazards. The two respirator classes (air purifying respirators and powered air purifying respirators) most commonly used for protection against particulates use filter media to collect/trap particles before they reach the user's breathing zone. Among the various test methods and criteria NIOSHuses as part of the certification process, respirator filter performance testing is the one most affected by the particle size. Since respirator users are exposed to a variety of hazards in different scenarios, respirator certification filtration testing was designed to use worst-case test conditions (e.g., different particle sizes and flow rates), so that filter performance in the workplace would not be worse. The NIOSH certification test for N-designated respirators uses a polydisperse distribution of NaCl particles with a count median diameter (CMD) of 0.075 +/-0.020 ѕm and a geometric standard deviation (GSD) of less than 1.86 [NIOSH 2005a]. NIOSH tests R- and P-designated respirators using a polydispersal of dioctyl phthalate (DOP) particles with a CMD of 0.185 +/-0.020 ѕm and a GSD of less than 1.60 [NIOSH 2005b]. For the lognormal distribution of NaCl aerosols used in the N series certification test, a broad range of particle sizes (e.g., 95% of the particles lie in the range of 22–259 nm) with a MMD of about 240 nm is used to determine whether the respirator filter performance is at least 95, 99, or 99.97% efficient. Most of the particles penetrating through the filter are measured simultaneously using a forward light scattering photometer. However, as noted in a recent review, the instrumentation used in the NIOSH certification test is not capable of measuring the light scattering of all particles less than 100 nm [Eninger et al. 2008a].

Particles larger than 0.3 *s*m are collected most efficiently by impaction, interception, andgravitational settling, while particles smaller than 0.3 μm are collected most efficiently by diffusion or electrostatic attraction [Hinds 1999]. In the development of the test method used for respirator certification, penetration by particles with an approximate 0.3 μm diameter was consideredto be the worst case because these particles were considered to be in the range of the most penetrating particle size [Stevens and Moyer 1989; TSI 2005; NIOSH 1996]. However, in practice, the most penetrating particle size range (MPPS) for a given respirator can vary based on the type of filter media employed and the condition of the respirator. For example, the most penetrating particle size for N95 air purifying respirators containing electrostatically charged filter media can range from 50–100 nm [Martin and Moyer 2000; Richardson et al. 2005] to 30–70 nm [Balazy et al. 2006; Eninger et al. 2008b]. These test results were recently confirmed by NIOSH [Rengasamy et al. 2007] in which five different models of respirators with N95 filters were challenged with 11 different monodisperse NaCl particles ranging in size from 20–400 nm. The monodisperse aerosol penetrations showed that the MPPS was in the 40-nm range for all respirator models tested. Under the aggressive laboratory test conditions employed in the study, mean penetration levels for 40-nm particles ranged from 1.4%–5.2%, which suggested that the respirators would be effective at capturing nanoparticles in the workplace. The NIOSH study also investigated whether there was a correlation between filtration performance using the existing NIOSH certification protocol for N series air purifying respirators and the filtration performance against monodisperse particles at the MPPS. A good correlation (r = 0.95) was found (e.g., respirators that performed better using the NIOSH certification test also had higher filter efficiencies against mono- disperse 40-nm nanoparticles), which is not surprising given that changes in filtration performance follow a consistent trend as a function of particle size.

According to single fiber filtration theory, below the most penetrating particle size, filtration efficiency will increase as particle size decreases. This trend will continue until the particles are so small that they behave like vapor molecules. As particles approach molecular size, they may be subject to thermal rebound effects, in which particles literally bounce through a filter. As a result, particle penetration will increase. The exact size at which thermal rebound will occur is unclear. However, a study by Heim et al. [2005] found that there was no discernable deviation from classical single-fiber theory for particles as small as 2.5-nm diameter. Subsequently, a NIOSH-funded contract with the University of Minnesota [Kim et al. 2007; Pui et al. 2006] and another study [Kim et al. 2006] showed that the penetration of nanoparticles through fibrous filter media decreased down to 2.5 nm as expected by the single fiber filtration theory. Thermal rebound phenomena were observed for nanoparticles below 2 nm diameter [Kim et al. 2006]. Recent studies provide additional data on nanoparticle penetration for NIOSH certified N95 and P100 filtering face-piece respirators [Rengasamy et al. 2008a], NIOSH certified N95 and European Certified FFP1 respirators [Huang et al. 2007], and FFP3 filter media [Golanski et al. 2008] using particles greater than 4 nm.

Based on these preliminary findings, NIOSH-certified respirators should provide the expected levels of protection if properly selected and fit tested as part of a complete respiratory protection program. However, as noted elsewhere [Rengsamy et al. 2007], in the unlikely event that the workplace exposure consists of a large percentage of particles in the most penetrating particle size range, management should take this information into account during the respiratorselection process, perhaps by choosing a respirator with higher levels of filtration performance (e.g., changing from an N95 to a P100, even though the APF will remain the same) as suggested by OSHA [Federal Register 2006] or by selecting a respirator with a higher APF (e.g., full face-piece respirator or powered air purifying respirator). Dust masks, commercially available at hardware/ home improvement stores, are often confused with NIOSH approved N95 filtering facepiece respirators because of their similar appearance. However, dust masks are not respirators and are not approved by NIOSH for respiratory protection. One study found that penetration of 40-nm NaCl nanoparticles range from 4.3%–81.6% for the seven dust mask models studied [Rengasamy et al. 2008b]. Dust masks should not be used in place of NIOSH-approved respirators for protection against nanoparticles.

NIOSH is continuing to study the protection afforded by NIOSH-certified respirators against emerging hazards such as engineered nanomaterials—including workplace-protection-factor studies—to ensure they provide expected levels of protection. NIOSH is also committed to updating 42 CFR Part 84—the regulatory language that provides NIOSH the authority to certify the performance of respirators in the United States—using a modular approach to rulemaking. Recently, NIOSH proposed the use of a TIL test as part of the respirator certification process for half-mask air purifying particulate respirators, including those having elastomeric and filtering face-pieces. The test protocol used to obtain benchmark TIL data for 101 half-face piece respirator models used 40–60 nm size ambient nanoparticles [NIOSH 2007]. Once implemented as part of the NIOSH certification process, the TIL tests should result inhalf-mask respirators with increased fitting performance. Future rulemaking activities may also include revisions to the filtration test to reflect changes in filtration performance resulting from use of new technologies (e.g., electret filter media). Results will be communicated as they become available.

8.3.6. Cleanup and disposal of nanomaterials

No specific guidance is currently available on cleaning up nanomaterial spills or contamination on surfaces; however, recommendations developed in the pharmaceutical industry for the handling and cleanup of pharmaceutical compounds might be applicable to worksites where engineered nanomaterials are manufactured or used [Wood 2001]. Until relevant information is available, it would be prudent to base strategies for dealing with spills and contaminated surfaces on current good practices, together with available information on exposure risks including the relative importance of different exposure routes. Standard approaches for cleaning powder spills include using HEPA-filtered vacuum cleaners, or wiping up the powder using damp cloths or wetting the powder prior to dry wiping. Liquid spills are typically cleaned by applying absorbent materials/liquid traps.

Damp cleaning methods with soaps or cleaning oils are preferred. Cleaning cloths should be properly disposed. Use of commercially available wet or electrostatic microfiber cleaning cloths may also be effective in removing particles from surfaces with minimaldispersion into the air. Drying and reusing contaminated cloths can result in re-dispersion of particles.

Energetic cleaning methods such as drysweeping or the using of compressed airshould be avoided or only used with precautions that assure that particles suspended by the cleaning action are trapped by HEPA filters. If vacuum cleaning is employed, care should be taken that HEPA filters are installed properly and bags and filters changed according to manufacturer's recommendations.

Table 8–2. Air-purifying particulate respirators

Respirator type	NIOSHassigned protectionfactor	Advantages	Disadvantages
Filtering facepiece (disposable)	10	• Lightweight • No maintenance or cleaning needed • No effect on mobility	• Provides no eye protection • Can add to heat burden • Inward leakage at gaps in face seal • Some do not have adjustable head straps • Difficult for a user to do a seal check • Level of protection varies greatlyamong models • Communication may be difficult • Fit testing required to select proper facepiece size • Some eyewear may interfere withthe fit
Elastomeric half-facepiece	10	• Low maintenance • Reusable facepiece and replaceable filters and cartridges • No effect on mobility	• Provides no eye protection • Can add to heat burden • Inward leakage at gaps in face seal • Communication may be difficult • Fit testing required to select proper facepiece size • Some eyewear may interfere withthe fit
Powered with loose-fitting facepiece	25	• Provides eye protection • Offers protection for people with beards, missing dentures or facial scars • Low breathing resistance • Flowing air creates cooling effect • Face seal leakage is generally outward • Fit testing is not required • Prescription glasses can be worn	• Added weight of battery and blower • Awkward for some tasks • Battery requires charging • Air flow must be tested with flowdevice before use

Table 8–2. (Continued)

Respirator type	NIOSH assigned protection factor	Advantages	Disadvantages
		• Communication easier than with elastomeric half-facepiece or full-facepiece respirators • Reusable components and replaceable filters	
Elastomeric full-facepiece with N-100, R-100, or P-100 filters	50	• Provides eye protection • Low maintenance • Reusable facepiece and replaceable filters and cartridges • No effect on mobility • More effective face seal than that of filtering facepiece or elastomeric half-facepiece respirators	• Can add to heat burden • Diminished field-of-vision compared to half-facepiece • Inward leakage at gaps in face seal • Fit testing required to select proper facepiece size • Facepiece lens can fog without nose cup or lens treatment • Spectacle kit needed for people who wear corrective glasses
Powered with tight-fitting half-facepiece or full-facepiece	50	• Provides eye protection with full-facepiece • Low breathing resistance • Face seal leakage is generally outward • Flowing air creates cooling effect • Reusable components and replaceable filters	• Added weight of battery and blower • Awkward for some tasks • No eye protection with half-facepiece • Fit testing required to select proper facepiece size • Battery requires charging • Communication may be difficult • Spectacle kit needed for people who wear corrective glasses with full face-piece respirators • Air flow must be tested with flow device before use

While vacuum cleaning may prove to be effective for many applications, the following issues should be considered. Forces of attraction may make it difficult to entrain particles off surfaces with a vacuum cleaner. The electrostatic charge on particles will cause them to be attracted to oppositely charged surfaces and repelled by similarly charged surfaces. A similarly charged vacuum brush or tool may repel particles, making it difficult to capture the aerosol or even causing it to be further dispersed. Vigorous scrubbing with a vacuum brush or tool or even the friction from high flow rates of material or air on the vacuum hose can generate a charge. The vacuum cleaners recommended for cleaning copier and printer toners have electrostatic-charge-neutralization features to address these issues.

When developing procedures for cleaning up nanomaterial spills or contaminated surfaces, consideration should be given to the potential for exposure during cleanup. Inhalation exposure and dermal exposure will likely present the greatest risks. Consideration will therefore need to be given to appropriate levels of personal protective equipment. Inhalation exposure in particular will be influenced by the likelihood of material reaerosolization. In this context, it is likely that a hierarchy of potential exposures will exist, with dusts presenting a greater inhalation exposure potential than liquids, and liquids in turn presenting a greater potential risk than encapsulated or immobilized nanomaterials and structures.

As in the case of any material spill or cleaning of contaminated surfaces, the handling and disposal of the waste material should follow existing federal, state, or local regulations

9. Occupational Health Surveillance

Occupational health surveillance is an essential component of an effective occupational safety and health program. The unique physical and chemical properties of nanomaterials, the increasing growth of nanotechnology in the workplace, and information suggesting that exposure to some engineered nanomaterials can cause adverse health effects in laboratory animals all support consideration of an occupational health surveillance program for workers potentially exposed to engineered nanomaterials [Schulte et al. 2008a]. Continued evaluation of toxicologic research and workers potentially exposed to engineered nanomaterials is needed to inform NIOSH and other groups regarding the appropriate components of occupational health surveillance for nanotechnology workers.

NIOSH has developed interim guidance relevant to medical screening (one component of an occupational health surveillance program) for nanotechnology workers (see NIOSH Current Intelligence Bulletin: *Interim Guidance on Medical Screening and Hazard Surveillance for Workers Potentially Exposed to Engineered Nanoparticles,* www. cdc.gov/niosh/review/public/115/). Medical screening is only part of a complete safety and health management program that follows the hierarchy of controls and involves various occupational health surveillance measures. Since specific medical screening of workers exposed to engineered nanoparticles has not been extensively discussed in the scientific literature, this document is intended to fill the knowledge gap on an interim basis.

Increasing evidence indicates that exposure to some engineered nanoparticles can cause adverse health effects in laboratory animals, but no studies of workers exposed to the few engineered nanoparticles tested in animals have been published. The current body of evidence about the possible health risks of occupational exposure to engineered nano- particles is quite small. Insufficient scientific and medical evidence now exists to recommend the specific medical screening of workers potentially exposed to engineered nanoparticles. Nonetheless, the lack of evidence on which to recommend specific medical screening does not preclude its consideration by employers interested in taking precautions beyond standard industrial hygiene measures [Schulte et al. 2008b]. If medical screening recommendations exist for chemical or bulk materials of which nanoparticles are composed, they would apply to nanoparticles as well.

Ongoing research on the hazards of engineered nanoparticles is needed along with the continual reassessment of available data to determine whether specific medical screening is warranted for workers who are producing or using nanoparticles. In the meantime, the following recommendations are provided for the management of workplaces where employees may be exposed to engineered nano- particles in the course of their work:

- Take prudent measure to control workers' exposures to nanoparticles.
- Conduct hazard surveillance as the basis for implementing controls.
- Continue use of established medical surveillance approaches.

NIOSH will continue to examine new research findings and update its recommendations about medical screening programsfor workers exposed to nanoparticles. Additionally, NIOSH is seeking comments on the strengths and weaknesses of exposure registries for workers potentially exposed toengineered nanoparticles.

10. Research Needs

NIOSH has developed a strategic plan for research on several occupational safety and health aspects of nanotechnology. The plan is available at www.cdc.gov/niosh/topics/nanotech/strat_plan.html. NIOSH has focused its research efforts in the following 10 critical topic areas to guide in addressing knowledge gaps, developing strategies, and providing recommendations.

1. **Exposure Assessment**
    - Determine key factors that influence the production, dispersion, accumulation, and re-entry of nano materials into the workplace.
    - Determine how possible exposures to nanomaterials differ by work process.
    - Assess possible exposure when nanomaterials are inhaled or settle on the skin.
2. **Toxicity and Internal Dose**
    - Investigate and determine the physical and chemical properties (e.g., size, shape, solubility, surface area, oxidant generation potential, surface functionalization, surface charge, chemical composition) that influence the potential toxicity of nanomaterials.
    - Determine the deposition pattern of nanoparticles in the lung and their translocation to the interstitium and to extrapulmonary organs.
    - Evaluate short- and long-term effects of pulmonary exposure to nanomaterials in various organ systems and tissues (e.g., lungs, brain, cardiovascular).
    - Determine if intratracheal instillation or pharyngeal aspiration can mimic the biological response to inhalation exposure to nano- materials.

- Determine the dermal effects of topical exposure to nano-objects, whether these nano-objects can penetrate into the skin, and whether they can cause immune alterations.
- Determine the genotoxic and carcinogenic potential of nano-objects.
- Determine biological mechanismsfor potential toxic effects.
- Determine whether in vitro screening tests can be predictive on in vivo response.
- Create and integrate models tohelp assess potential hazards.
- Determine whether a measure other than mass is more appropriate for determining toxicity.

3. **Epidemiology and Surveillance**
   - Evaluate existing exposure and health data for workers employed in workplaces where nanomaterials are produced and used, with emphasis on improving our understanding of the value andutility of establishing exposure registries for workers potentially exposed to engineered nanomaterials.
   - Assess the feasibility of industrywide exposure and epidemiological studies of workers exposed to engineered nanomaterials, with emphasis on workers potentially exposed to engineered carbonaceous nanomaterials.
   - Integrate nanotechnology safety and health issues into existing hazard surveillance mechanisms and continue reassessing guidance related to occupational health surveillance for workers potentially exposed to engineered nanomaterials.
   - Build on existing public health geographical information systems and infrastructure to enable effective and economic development of methods for sharing nanotechnology safety and health data.
4. **Risk Assessment**
   - Determine how existing exposure-response data for fine and ultrafine particles (human or animal) may be used to identify the potential hazards and estimate the potential risks of occupational exposure to nanomaterials.
   - Develop a framework for assessing the potential hazards and risks of occupational exposure to nano- materials, using new toxicologic data on engineered nanomaterials and standard risk assessment models and methods.
5. **Measurement Methods**
   - Evaluate methods used to measure the mass of respirable particles in the air and determine whether this measurement can be used to measure nanomaterials.
   - Develop and field-test practical methods to accurately measure airborne nanomaterials in the workplace.
   - Develop, test, and evaluate systems to compare and validate sampling.
6. **Engineering Controls and Personal Protective Equipment**
   - Evaluate the effectiveness of engineering controls in reducing occupational exposures to nano- aerosols and developing new controls when needed.
   - Evaluate the suitability of control- banding techniques when additional information is needed and evaluate the effectiveness of alternative materials.
   - Evaluate and improve current personal protective equipment.
   - Develop recommendations (e.g., use of respiratory protection) to prevent or limit occupational exposures to nanomaterials.

7. **Fire and Explosion Safety**
    - Identify physical and chemical properties that contribute to dustiness, combustibility, flammability, and conductivity of nanomaterials.
    - Recommend alternative work practices to eliminate or reduce work place exposures to nanomaterials.
8. **Recommendations and Guidance**
    - Use the best available science to make interim recommendations for workplace safety and health practices during the production, use, and handling of nanomaterials.
    - Evaluate and update mass-based occupational exposure limits for airborne particles to ensure good, continuing precautionary practices.
9. **Communication and Information**
    - Establish partnerships to allow for identification and sharing ofresearch needs, approaches, and results.
    - Develop and disseminate training and education materials to workers, employers, and occupational safety and health professionals.
10. **Applications**
    - Identify uses of nanotechnology for application in occupational safety and health.
    - Evaluate and disseminate effective applications to workers, employers, and occupational safety and health professionals.

Appendix. Nanoparticle Emission Assessment Technique for Identification of Sources and Releases of Engineered Nanomaterials

1.0. Introduction

This appendix describes a technique that can be used by industrial hygienists for conducting initial workplace assessments for possible nanoparticle emissions. It allows a semi-quantitative evaluation of processes and tasks in the workplace where releases of engineered nanoparticles may occur. NIOSH uses several sampling approaches simultaneously with the goal of obtaining key physicochemical particle metrics: number concentration, qualitative size, shape, degree of agglomeration, and mass concentration of elemental constituents of interest.

2.0. Scope

Employers, workers, and researchers engaged in the production and use of engineered nanomaterials have expressed an interest in determining whether these nano- materials are hazardous and if the potential for worker exposure exists. NIOSH has an active toxicology program to assess the potential hazards of engineered nanoparticles. Unfortunately these studies require long time periods and fall behind the pace of production and use of these nanomaterials. To assist in answering the latter of these questions, NIOSH established a nanotechnology field research team tasked with visiting facilities and collecting information about the potential for release of nanomaterials and worker exposure at those facilities. The initial challenges that the field research team encountered were: 1) determining which exposure metric (e.g., mass, particle number concentration, particle surface area) for engineered nanoparticles would provide a consistent body of knowledge to align with the toxicological results observed in experimental animal studies; and 2) selecting a sampling method based on metrics that were practical and would provide reproducible results. Engineered nanomaterials can be measured in the workplace using a variety of instrumentation including: condensation particle counter (CPC); optical particle counter (OPC); scanning mobility particle sizer (SMPS); electric low pressure impactor (ELPI); aerosol diffusion charger; and tapered element oscillating microbalance (TOEM), which vary in complexity and field portability. Unfortunately, relatively few of the above instruments are readily applicable to routine exposure monitoring due to non-specificity, lack of portability, difficulty of use, and high cost. NIOSH researchers have developed and used a field assessment strategy for determining exposures to engineered nanoparticles that could be adopted by other health and safety professionals in the evaluation of occupational exposures [Methner, et. al. 2007; Methner, 2008].

Since there are currently no exposure limits specific to engineered nanomaterials, this technique is used to determine whether airborne releases of engineered nanomaterials occur. This assessment, which compares particle number concentrations and relative particle size at the potential emission source to background particle number concentrations and particle size, provides a semi- quantitative means for determining the effectiveness of existing control measures in reducing engineered nanoparticle exposures.This procedure utilizes portable direct-reading instrumentation supplemented by filter- based air samples (source-specific and personal breathing zone [PBZ]). The use of filter samples is crucial for particle identification because direct-reading instruments used for determining particle number concentrations are incapable of identifying the composition of the particles.

3.0. Summary of the On-Site Initial Assessment

The initial assessment uses a combination of direct-reading, handheld instruments (CPC and OPC) and filter-based sampling (e.g. 37-mm diameter filter cassettes) for subsequent chemical and microscopic analyses (Figure 1). This semi-quantitative approach was first described by Maynard et al. [2004] and NIOSH has adopted a similar approach. The technique includes determining particle number concentration using direct-reading, handheld particle counters at potential emission sources and comparing those data to background particle number concentrations. If elevated concentrations of suspected nanoparticles are detected at potential emission sources, relative to the background particle number concentrations, then a pair of filter-based, source-specific air samples are collected with one sample analyzed by transmission electron microscopy (TEM) or scanning electron microscopy (SEM) for particle identification and characterization, and the other used for determining the elemental mass concentration (Figure 2). A second pair of filter-based air samples may also be collected in the personal breathing zone of workers. Breathing zone samples are analyzed in the same manner as the area air samples (i.e., by TEM and elemental mass).

Figure 1. A demonstration of the initial assessment technique with side-by-side sampling using (from left to right) the OPC, co-located open-face filter cassettes, and the CPC: examples of PBZ and source-specific filter-based sampling setup.

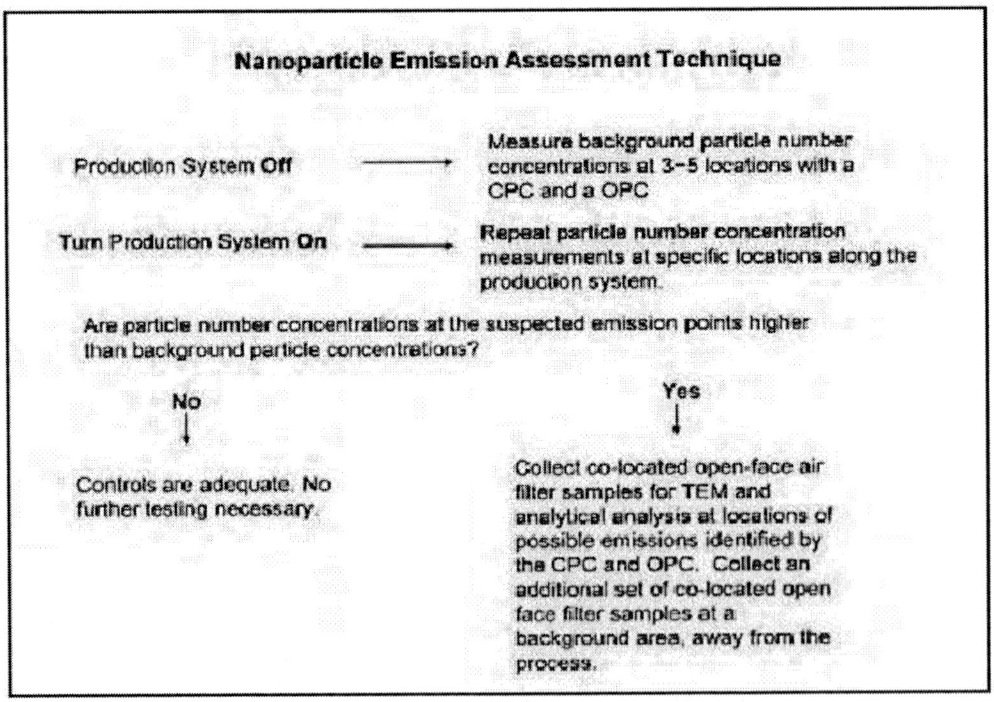

Figure 2. Summary of the initial assessment technique

4.0. Air Sampling Instrumentation and Filter Media Used in the Initial Assessment

The following instrumentation is used by NIOSH; however, use does not constitute endorsement.

4.1 TSI model 3007 (or model 8525) (TSI Inc, Shoreview, MN), handheld condensation particle counter (CPC), which uses isopropanol to condense on particles so they can be counted
The TSI units provide a non-specific measure of the total number of particles independent of chemical identity per cubic centimeter of air (P/cm^3). The measureable range is between 10–1,000 nm for model 3007, or between 20–1,000 nm for model 8525. The range of detection for these instruments is reported by the manufacturer to be 0–100,000 P/cm^3.

4.2 ART Instruments Hand Held Particle Counter (HHPC-6, ART Instruments, Grants Pass, Oregon), which operates on optical counting principles using laser light scattering.
The HHPC-6 optical particle counter (OPC) can measure the total number of particles per liter (P/L) of air independent of chemical identity within six specific size ranges. The OPC used by the NIOSH field research team provides particle counts in the following size cut- points: 300 nm; 500 nm; 1,000 nm; 3,000 nm; 5,000 nm; and 10,000 nm. The range of detection for this instrument is reported by the manufacturer to be 0–70,000 P/L. Different manufacturers' OPCs may have slightly different particle size ranges and could be substituted.

4.3 Appropriate air sampling filter media (e.g. mixed cellulose ester, quartz fiber filter) are selected depending on nanoparticle type and desired analytical information (e.g., determination of particle morphology using TEM or SEM, elemental analysis for metals, elemental analysis for carbon)

4.4 Air sampling pumps capable of sampling at high flow rates (e.g., 7 liters per minute or other flow rate depending upon the duration of the task and the appropriate NIOSH method, if a method is available)

4.5 Sampling pump flow calibrator

4.6 If desired, personal cascade impactor or respirable cyclone (see 5.3.3)

4.7 If desired, cassette conductive cowl (see 5.3.3)

4.8 Optional research-grade particle analyzers for expanded surveys (see 5.6.1)

4.9 Optional surface sampling supplies such as substrate (e.g., Ghost Wipes™), disposable 10 cm × 10 cm templates, sterile containers, and nitrile gloves for handling media (see 5.6.2)

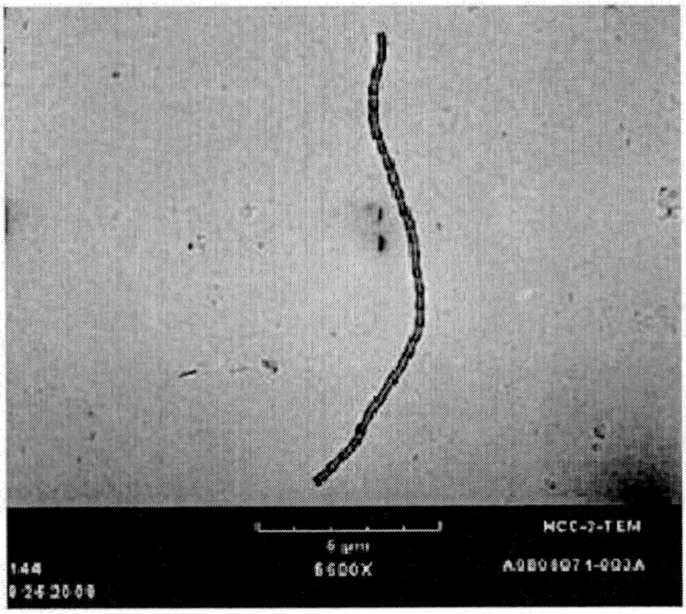

Figure 3. Electron microscopy micrograph of a carbon nanofiber

5.0. Evaluation of Potential Releases of Engineered Nanomaterials

5.1. Identify Potential Sources of Emissions

The overall purpose of this step is to develop a list of target areas and tasks that will be evaluated with the particle analyzers.

The initial assessment involves identifying the potential sources of engineered nanomaterial emissions by reviewing the type of process, process flow, material inputs and discharges, tasks, and work practices. When available, literature (e.g., MSDS, records of feedstock materials) is reviewed to gain an understanding of the engineered nanomaterials being produced or used, including their physicochemical properties such as size, shape, solubility, and reactivity. Once the potential sources of emissions have been identified from the process review, the industrial hygienist (or other qualified person):

- Conducts an observational walk- through survey of the production area and processes to locate potential sources of emissions.
- Determines the frequency and duration of each operation and the type of equipment used for handling and containment of the material.
- Determines the presence/absence of general and local exhaust ventilation and other engineering controls. (This initial assessment includes identifying points of potential system failure that could result in emission from the containment/control system [e.g., hole in duct, deteriorated sealing gasket]).
- Determines the process points where containment is deliberately breached (e.g., opening system for product retrieval or for cleaning).

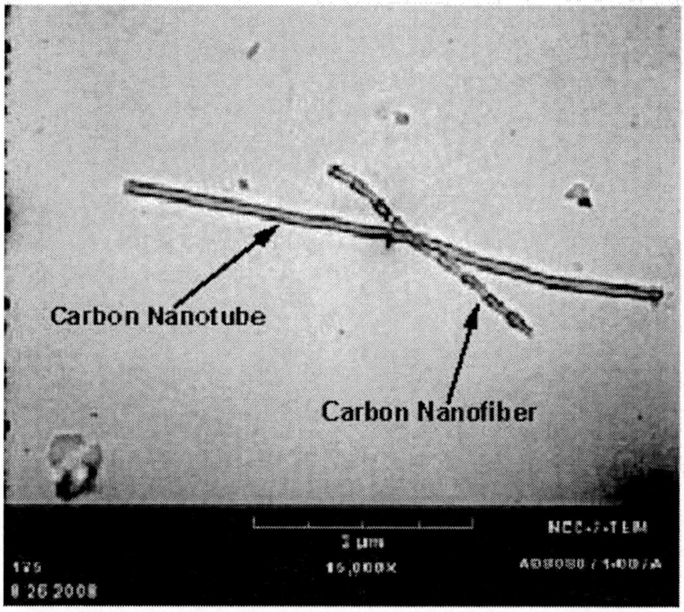

Figure 4. Electron microscopy micrograph of a carbon nanofiber and carbon nanotube

5.2. Conduct Particle Concentration Sampling

5.2.1. Background measurements

Determining the contribution of background particle concentrations on measurements made for the particles of interest (e.g., engineered nano- particles) is an important evaluation of assessing the possible airborne release of engineered nanoparticles.

Ideally, during the initial assessment, the industrial hygienist (or other qualified person), will determine the average airborne particle concentration at various processes and adjacent work areas with the CPC and OPC *before* the processing or handling of nano- materials begins. If the background particle concentrations are high (values are relative and will vary with processes and facilities), an assessment is made as to whether there may be a source of incidental nanoparticles in the area. Incidental nanoparticles may be generated from a variety of sources, including vacuum pumps, natural gas heating units, gasoline/propane/diesel powered fork lift trucks, or other combustion activities such as welding, soldering, or heat-sealing. The CPC and OPC can be used to check these sources for incidental nanoparticle releases. Outdoor or re-circulated air supply from the building ventilation system should also be considered as a possible source of nanoparticles [Peters et al. 2006].

Measurements of background particle concentrations are repeated after the active processing, manufacturing, or handling of the nanomaterial has ended. An average background concentration is then computed and subtracted from the measurements made during processing, manufacturing, or the handling of engineered nanomaterials. This approach is acceptable only if background particle counts remain relatively stable throughout the measurement period and particle emissions from the process under investigation are sufficiently elevated above background. For other situations, correcting for particle back-ground concentrations becomes more complex requiring additional sampling over an extended time period to determine the source(s) and magnitude of background particle concentrations. This type of evaluation is generally outside the scope of the initial assessment described here.

5.2.2. Area sampling

Once initial background particle concentrations have been determined, measurements of airborne particle concentrations and size ranges are made with the CPC and OPC simultaneously at locations near the suspected or likely emission source (e.g., opening a reactor, handling product, potential leak points in the ventilation system). Airborne particle concentrations are determined before, during, and after each task or operation to identify those factors (e.g., controls, worker interaction, work practices) that may affect airborne particle concentrations. This information is used to identify processes, locations, and personnel for filter- based air sampling (5.3).

5.3. Conduct Filter-based Area and Personal Air Sampling

5.3.1. Area air sampling

A pair of filter-based, air samples are collected at process/task locations and/or workers engaged in process operations where suspected engineered nanomaterial emissions may occur, based on air sampling results using the CPC and OPC.

Filter-based area air samples provide more specific information on the engineered nanomaterial of interest (e.g., size, shape, mass). The pair of air samples includes one sample analyzed for elemental mass and one sample analyzed by electron microscopy. For example, one sample might be collected for metals determination (e.g., NIOSH Method 7300, 7303) or elemental carbon (e.g., NIOSH Method 5040) depending on the composition of the engineered nano- material. The other sample would be collected for particle characterization (e.g., size, shape, dimension, degree of agglomeration) by TEM or SEM using the measurement techniques specified in NIOSH Methods 7402, 7404, or other equivalent methods [NIOSH 1994].

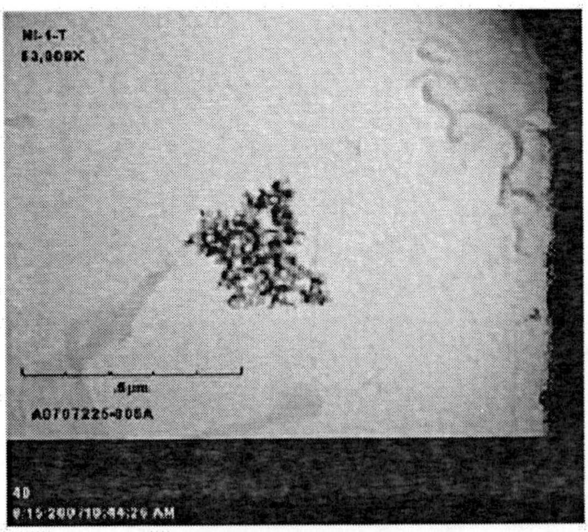

Figure 5. Electron microscopy micrograph of an agglomerated nanoparticle of nickel oxide

The source-specific air samples are collected as close as possible to the suspected emission source but outside of any existing containment, to increase the probability of detecting any possible release of engineered nanomaterials. Sampling duration generally matches the length of time in which the potential exposure to the engineered nanomaterial exists at the task or specific process. In cases where the duration of the tasks associated with the potential airborne release of nanomaterials is short (e.g., minutes), a relatively high air sampling flow rate may be required (approximately 7 liters per minute) to ensure adequate particle loading on the filter media. If specific information is desired on the worker's potential exposure to the engineered nanomaterial then PBZ samples should be collected using the two-sample filter-based sampling strategy described above.

If the particle number concentrations (using CPC or OPC) are substantially high, then shorter sampling times for the TEM or SEM sample may be necessary to avoid overloading the filter and interfering with particle characterization. The specific sampling time should be based on direct-reading instrument results and professional judgment of the industrial hygienist. In general, filter samples are collected for the duration of a given task, normally 15–30 minutes. If the direct-reading instruments indicate a high particle number concentration the sampling time can be shortened to 5–10 minutes, or both a short- and long-duration sample may be collected to ensure an adequate sample for electron microscopy analysis. See Table 1 for additional sampling time guidance. However, the sampling times in Table 1 were based on collection of asbestos fibers by NIOSH Method 7402 and may not be applicable for much smaller engineered nanoparticles. See Figures 3–5 for example TEM micrographs.

Table 1. Approximate Sampling Times for TEM Grid Based on Particle Number Concentrations*

	Open-faced cassettes			
	TEM grid	**25-mm**	**37-mm**	**47-mm**
Diameter (mm)	3.0	25.0	37.0	47.0
Effective diameter (mm)	3.0	22.2	34.2	44.2
Effective collection area (mm^2)	7	385	916	1531
Flow (L/min)	0.1	7	7	7
Desired Loading (#/mm^2)	1.E+06	1.E+06	1.E+06	1.E+06
Air concentration (#/cm^3)	**Time (min)**			
250	282.7	220.2	523.4	874.8
500	141.4	110.1	261.7	437.4
1,000	70.7	55.0	130.8	218.7
2,000	35.3	27.5	65.4	109.4
4,000	17.7	13.8	32.7	54.7
8,000	8.8	6.9	16.4	27.3
16,000	4.4	3.4	8.2	13.7
32,000	2.2	1.7	4.1	6.8
64,000	1.1	0.9	2.0	3.4
128,000	0.6	0.4	1.0	1.7

*NIOSH NMAM Method 7402 Asbestos by TEM and personal communication with Dr. Aleksandr Stefaniak (NIOSH)

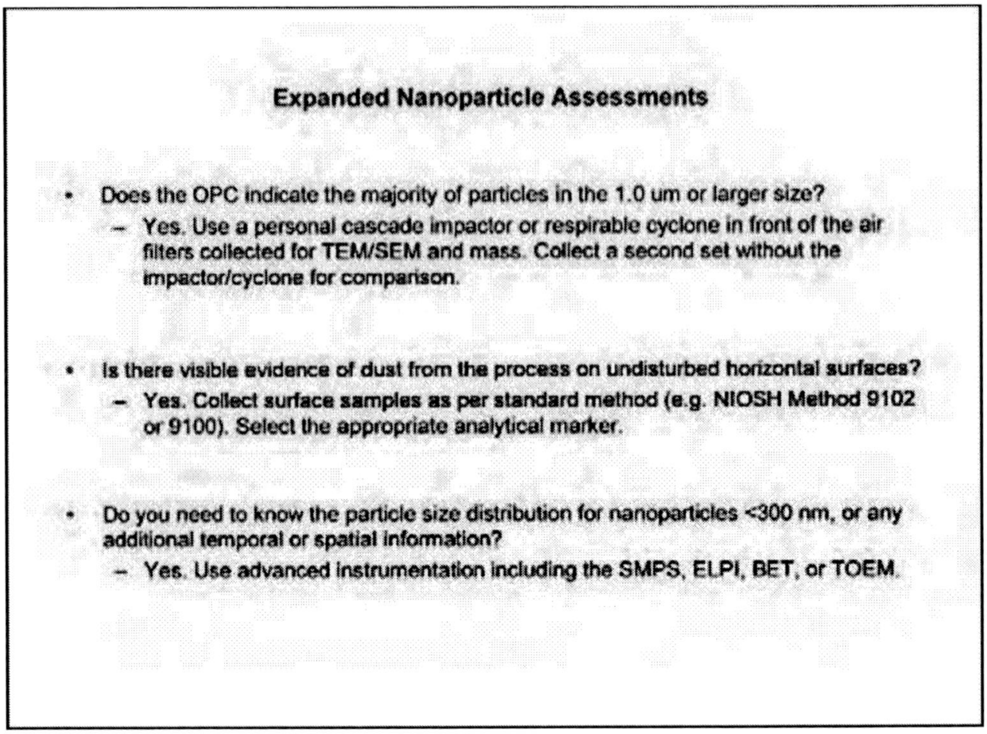

Figure 6. Considerations for expanded nanomaterial assessments

A minimum of 2 background filter samples are collected distant from the potential sources of engineered nanoparticle exposure to serve as an indicator of ambient particle identification and concentration.

5.3.2. Personal air samples

When possible, personal breathing zone (PBZ) air samples are collected on workers likely to be exposed to engineered nanomaterials (e.g., engaged in active handling of nanomaterials or operating equipment previously identified as emitting nanoparticles). If measurements obtained with the CPC and OPC indicate that nano- particles are being emitted at a specific process where a worker is located, then the collection of PBZ samples may be warranted.

PBZ samples are analyzed in the same manner as the area air samples (i.e., by TEM and elemental mass). It may be necessary to collect samples at a relatively high flow rate (e.g., 7 liters per minute) if the duration of the task and the resulting potential exposure is short.

5.3.3. Optional sample collection

In the event that measurements made by the OPC indicate a large fraction (over 50%) of particles exceeding 1,000 nm in size, the use of a personal cascade impactor or respirable cyclone sampler in tandem with a filter-based air sampling cassette may be required for both the mass and TEM/SEM analyses to eliminate large particles that may interfere with analysis and be of limited interest. The use of an impactor or cyclone will require using a flow rate appropriate for the particle cut size and is usually in the range of 1.7–2.5 liters per minute. Open-face, and impactor or cyclone samples may be collected side by side to allow a more thorough interpretation of analytical results. Additionally, if it is anticipated that the nanoparticles of interest will have a tendency to be electrostatically attracted to the sides of the plastic air sampling cassette, a conductive cowl may be necessary to eliminate particle loss and subsequent underestimation of the airborne nanoparticle concentration. The use of a personal cascade impactor, respirable cyclone, or conductive cowl is made at the discretion of the industrial hygienist (or other qualified person).

If the facility is manufacturing or using TiO_2, then the sampling should include the sampling recommendations found in the NIOSH *Draft Document: Evaluation of Health Hazard and Recommendations for Occupational Exposure to Titanium Dioxide* (www.cdc.gov/niosh/review/public/TiO2/default. html),which recommends collecting a mass-based airborne measurement using NIOSH Method 0600.

5.4. Quality Assurance and Quality Control

To ensure valid emission measurements, the following quality assurance and control steps should be taken:

- Use factory calibrated direct- reading particle analyzers
- Perform daily zero-checks on all particle counters before each use
- Calibrate pumps before and after each sampling day
- Submit for analysis any process, background, and bulk material samples along with field and media blanks to a laboratory accredited by the American Industrial Hygiene Association (AIHA)

5.5. Data Interpretation

Since the size of airborne engineered nanoparticles and the degree of agglomeration may be unknown at the time of sample collection, the use of direct-reading, particle sizing/ counting instruments may provide a semi-quantitative indication of the magnitude of potential emissions, provided background particle number subtraction can be successfully accomplished. The particle number concentration measurements taken with the CPC and OPC will provide a measurement of particles larger than the ASTM definition of nanoparticles (1–100 nm) [ASTM 2006]. However, the two particle counters can be used simultaneously to obtain a semi- quantitative size differential evaluation of the aerosol being sampled. The CPC provides a measure of total particles per cm^3 in the size range of 10–1,000 nm (or 20–1,000 nm). The OPC provides the total number of particles per liter of air within six specific size ranges: 300 nm; 500 nm; 1,000 nm, 3,000 nm, 5,000 nm and > 10,000 nm. If necessary, the data from the CPC and OPC can be used together to determine the number concentration of nanoscale particles. For example, a high particle number concentration on the CPC, in combination with a high particle number concentration in the small size ranges (300–500 nm) on the OPC, may indicate the possible presence of nanoscale particles. Conversely, a low CPC particle number concentration, in combination with a high OPC particle number concentration in the larger size ranges (> 1,000 nm) may indicate the presence of larger particles and/or engineered nanoparticle agglomerates. These assumptions of nanoparticles versus larger particles and/or nanoparticle agglomerates may be verified by TEM or SEM analysis.

5.5.1. Selectivity

Selectivity is a critical issue when characterizing exposure using airborne particle number concentration. Airborne nanoparticles are present in many workplaces and often originate from multiple sources such as combustion, vehicle emissions, and infiltration of outside air. Particle counters are generally not selective to particle source or composition, making it difficult to differentiate between incidental and process-related nano- particles using number concentration alone. The CPC and OPC are used to identify sources of nanoparticles and the filter-based samples are used to verify the size, shape, and chemical composition of the nanoparticles with the goal of differentiating between incidental and engineered nanoparticles.

5.5.2. Limitations

The exposure assessment technique does have some limitations including:

- Although this issue is not unique to particle number concentration measurements, orders of magnitude difference can exist in aerosol number concentrations, depending on the number and types of sources of particle emissions. Monitoring over several days and during different seasons can provide a better understanding of the variability that might exist in airborne particle number concentrations found in background measurements and in measurements made at sources where engineered nano- materials are handled.
- The upper dynamic range of the CPC is 100,000 P/cm^3. A dilutor, consisting of a modified HEPA filter cartridge placed upstream of the inlet, can extend the range of the CPC when particle number concentrations are greater than 100,000 P/cm^3 [Peters et al. 2006; Heitbrink et al. 2007; Evans et al. 2008].
- The analysis of air samples by TEM or SEM with energy dispersive X-ray spectrometry can provide information on the elemental composition of the nanomaterials. However, TEM and SEM analysis can be compromised if there is particle overload on the filter. Alternatively, if the loading is too sparse, an accurate assessment of particle characteristics may not be possible (see 5.3.1).
- Note that area samples are collected as closely as possible to the source of emission to allow for more accurate determination of a nanoparticle release and to identify locations most likely to result in worker exposure.Therefore, results from this type of sampling should not be interpreted as representative of worker exposure. However, samples collected in such a fashion should serve as an indicator of material release and the possible need for controls.

5.6. Expanded Research (In Depth Assessments)

5.6.1. Research instrumentation

A major obstacle in conducting more specific measurement of engineered nanomaterials in the workplace is a lack of field-portable instruments that can be easily maneuvered within a facility or easily worn by a worker to provide an indication of PBZ exposure. Additionally, there is no single instrument capable of measuring the numerous potential exposure metrics associated with engineered nanomaterials (e.g., number concentration, surface area, size, shape, mass concentration) [Maynard and Aitken 2007]. Although the following instruments lack field portability and ease of use, they can measure many of the desirable exposure metrics and provide information about the particle size distribution. These research-grade particle analyzers are not usually part of the initial assessment but are used when additional knowledge about the nanoscale particle temporal or spatial exposure variation or size distribution is desired.

5.6.1.1. Particle Surface-Area Analyzers

Toxicology studies have indicated that surface area of nanoparticles may be an important exposure dose metric. Portable aerosol diffusion chargers may be used to provide estimates of external aerosol surface area when airborne particles are smaller than 100 nm in diameter, but these may tend to overestimate external surface area when particles are larger than 100 nm in diameter. These instruments are based on diffusion charging followed by detection of the charged aerosol using an electrometer.

The TSI Aerotrak™ 9000 Nanoparticle Aerosol Monitor does not measure total active surface area but indicates the surface area of particles which may be deposited in the lung in units of square micrometers per cubic centimeter, corresponding to either the tracheobronchial or alveolar regions of the lung. The Ecochem DC 2000-CE measures the total particle surface area. These devices are currently being evaluated as part of the process used by NIOSH to conduct initial assessments. These particle surface analyzers are used as area samplers.

5.6.1.2. Scanning Mobility Particle Sizer

More specific depictions of particles by size (diameter) and number can greatly improve the ability to evaluate possible releases of engineered nano- particles. One particular instrument, the Scanning Mobility Particle Sizer (SMPS) measures particle diameters from 2.5–1,000 nm and can display data as a size and number distribution using up to 167 size channels. The SMPS is widely used as a research tool for characterizing nano- scale aerosols. The SMPS employs a continuous, fast-scanning technique to provide high-resolution measurements. However, the SMPS may take 2–3 minutes to scan which may not be useful for the process screening in workplaces with highly variable aerosol size distributions. Its applicability for use in the workplace may be limited because of its size, cost, and use of an internal radioactive source.

The Fast Mobility Particle Sizer (FMPS) is similar to the SMPS but has a much faster response time (approximately 1 second). However, because it has fewer particle size channels, it does not include the same level of detail on particle size distributions that can be determined with the SMPS.

The FMPS and SMPS are used as area samplers.

5.6.1.3. Low Pressure Impactors

The Electrical Low Pressure Impactor (ELPI) combines diffusion charging and a cascade impactor to provide aerosol size distributions by aerodynamic diameter as determined real time by mass and number collected on a series of plates.

Low pressure cascade impactors offer the ability to size particles and then conduct secondary analyses (e.g., metals analysis). However, these instruments are sensitive to harsh field conditions and are not considered portable. The ELPI is used as an area sampler.

5.6.1.4. Tapered Element Oscillating Microbalance

The tapered element oscillating microbalance (TEOM) is commonly used for sampling aerosols less than 1 μm in diameter, however, the sampling inlet can be set to select different size fractions. The TEOM determines mass by detecting a change in vibration frequency across a particle-collecting substrate. The TEOM can be configured to provide size-differentiated mass measurements and is used as an area sampler.

5.6.2. Surface sampling

Surface sampling to detect the presence of engineered nanomaterials is not routinely part of the initial assessment but may be conducted to determine if surface contamination exists. Surface sampling does not provide size-specific information but may be useful for determining whether engineered nanomaterials have migrated away from active production or handling areas and have contaminated nonproduction work areas. The decision to collect surface samples is made in the field at the discretion of the industrial hygienist (or other qualified person), and is dependent on direct observation and the nano- material of interest. For example, surface sampling was completed at a quantum dot facility after observing dusty surfaces in areas adjacent to the production area. In order to determine if the dust was contaminated with quantum dots, surface samples were collected and analyzed for the chemical components of the quantum dots produced by that facility.

Surface wipe samples are collected using a pre-moistened substrate such as Ghost Wipe^{TM} towelettes in accordance with NIOSH Method 9102 for elements or the NIOSH method for specific elements (e.g., NIOSH Method 9100 for lead). When collecting wipe samples, the following steps should be followed:

- Don a pair of nitrile disposable gloves
- Wipe the surface within a disposable 10 cc × 10 cc template using four horizontal s-shaped strokes
- Fold the exposed side of the wipe in and wiping the same area with four vertical s-shaped strokes
- Fold the wipe, exposed side in, and placing it into a sterile container Gloves and template are discarded after each sample collection to eliminate the possibility of cross- contaminating successive samples. Wipe samples may be collected from undisturbed horizontal surfaces throughout the facility at locations suspected to be contaminated and in areas expected to be free of engineered nanomaterials. Wipe samples are analyzed following the appropriate NIOSH method for the chemical substance of interest.

6.0. Conclusions

The NIOSH initial assessment technique uses complimentary approaches to semi-quantitatively evaluate the potential releases of engineered nanoparticles. Two different particle counters are used in a parallel and differential manner to evaluate the total particle number relative to background and the relative size distribution of the particles. If this initial evaluation indicates an elevated number of small particles, which could potentially be the engineered nanoparticle of interest, then the particle counters are used to detect the source of the emissions. If nanoparticles are found and determined to be emitted from a specific process (versus background incidental nanoscale particles), then additional samples are collected for qualitative measurement of particle size and shape, (by TEM or SEM analysis) and for determination of elemental mass concentration (by chemical analysis).

The initial assessment technique is useful for determining whether airborne releases of engineered nanomaterials are occurring at potential emission sources. This assessment provides a semi-quantitative means for determining whether existing measures are adequate for controlling nanomaterial emissions or if additional controls may be required.

The NIOSH emission assessment technique may be useful to health and safety professionals who are interested in determining whether release of nanomaterials occurs in the workplace. Where possible, use of the technique should be repeated in workplaces of interest to gain a better understanding of the daily fluctuations in airborne exposures at processes and tasks in which engineered nanomaterials occur and for determining potential sources of background particle number concentrations. A more systematic and routine assessment of the workplace can provide more definitive information on the performance of control measures and if additional actions are needed to reduce worker exposure.

The initial assessment technique can be expanded or modified to determine additional metrics (Figure 6). Research initiatives addressing more comprehensive process monitoring, particle metrics, personal exposure monitoring, and method/approach development and validation are currently underway within NIOSH. As this information becomes available, revisions to the Approaches to Safe Nanotechnology document will be made.

Information about contacting the nanotechnology field research teamis available at: [www.cdc.gov/niosh/docs/2008-121], see the Fact Sheet: NIOSH Nanotechnology Field Research Effort [NIOSH 2008].

7.0. References

ASTM International. (2006) *ASTME2456–06 Standard terminology relating to nanotechnology.*

Evans, D. E., Heitbrink, W. A., Slavin, T. J. & Peters, T. M. (2008). Ultrafine and respirable particles in an automotive grey iron foundry. *Ann Occup Hyg,52(1)*, 9–21.

Heitbrink, W. A., Evans, D. E., Peters, T. M. & Slavin, T. J. (2007). The characterization and mapping of very fine particles in an engine machining and assembly facility. *J Occup Environ Hyg, 4*, 341–351.

Maynard, A. & Aitken, R. (2007). Assessing exposure to airborne nanomaterials: current abilities and future requirements. *Nanotoxicology,1(1)*, 26–41.

Maynard, A., Baron, P., Foley, M., Shvedova, A., Kisin, E. & Castranova, V. (2004) Exposure to carbon nanotube material: Aerosol release during the handling of unrefined single walled carbon nanotubes material. J Toxicology. *Environmental Health Part A*, *67*, 1, 87–107.

Methner, M. (2008). Engineering case reports. Old L, ed. Effectiveness of local exhaust ventilation in controlling engineered nanomaterial emissions during reactor cleanout operations. *J Occup Environ Hyg,5(6)*, D63–D69.

Methner, M. M., Birch, M. E., Evans, D. E., Ku, B. K., Crouch, K. G. & Hoover, M. D. (2007). Mazzukelli LF, ed: Case study: Identification and characterization of potential sources of worker exposure to carbon nanofibers during polymer composite laboratory operations. *J Occup Environ Hyg,4(12)*, D125–D130.

NIOSH. (1994). NIOSH manual of analytical methods (NMAM®), 4th ed. By Schlecht PC, O'Conner PF, eds. Cincinnati, OH: U.S. Department of Health and Human Services, Centers for Disease Control and Prevention, National Institute for Occupational Safety and Health. DHHS (NIOSH) Publication No. 94–113 (www.cdc. gov/niosh/nmam/).

NIOSH. (2008). NIOSH Fact Sheet: NIOSH Nanotechnology Field Research Effort, Cincinnati, OH: U.S. Department of Health and Human Services, Centers for Disease Control and Prevention, National Institute for Occupational Safety and Health. DHHS (NIOSH) Publication No. 2008–12 1 (http://www.cdc.gov/ niosh/docs/2008- 121/).

Peters, T., Heitbrink, W., Evans, D., Slavin, T. & Maynard, A. (2006). The mapping of fine and ultrafine particle concentrations in an engine machining and assembly facility. *Ann Occup Hyg,50(3)*, 1–9.

References

71 Fed. Reg. 50121–50192. (2006). Occupational Safety and Health Administration: assigned protection factors, final rule (www.osha.gov/pls/oshaweb/ owadisp.show_document? p_table=FEDERAL_ REGISTER&pjd=1 8846).

ACGIH. (2001). *Industrial ventilation: a manual of recommended practice*. Cincinnati, OH: American Conference of Governmental Industrial Hygienists.

Adams R. J. & Bray, D. (1983). Rapid transport of foreign particles microinjected into crab axons. *Nature,303*,718–720.

Ambroise, D., Wild, P. & Moulin, J. J. (2007). Update of a meta-analysis on lung cancer and welding. *Scand J Work Environ Health, 32(1)*, 22–31.

Antonini, J. M. (2003). Health effects of welding. *Crit Rev Toxicol, 33(1)*, 61–103.

ASTM Committee E2535–07. (2007). *Standard guide for handling unbound engineered nanoscale particles in occupational settings*. West Conshohocken, PA: ASTM International.

ASTM Subcommittee F23.40 (2003). Standard test method for resistance of materials used in protective clothing to penetration by blood-borne pathogens using Phi–X174 bacteriophage penetration as a test system. West Conshohocken, PA: American Society for Testing and Materials, ASTM F1671–03.

Balazy, A., Toivola, M., Reponen, T., Podgorski, A., Zimmer, A. & Grinshpun, S. A. (2006). Manikin-based filtration performance evaluation of filtering-facepiece respirators challenged with nanoparticles.*Ann Occup Hyg, 50(3)*, 259–269.

Baltensperger, U., Gaggeler, H. W. & Jost, D. T. (1988). The epiphaniometer, a new device for continuous aerosol monitoring. *J Aerosol Sci, 19(7),* 931–934.

Barlow, P. G., Clouter-Baker, A. C., Donaldson, K., Mac- Callum, J. & Stone, V. (2005). Carbon black nanoparticles induce type II epithelial cells to release chemotaxins for alveolar macrophages. *Part Fiber Toxicol, 2(14)*.

Baron, P. A., Deye, G. J., Chen, B., Schwegler-Berry, D., Shvedova, A. A. & Castranova, V. [in press]. Aerosolization of single-walled carbon nanotubes for an inhalation study. Inhal Toxicol.

Bergman, W., Garr, J. & Fearon, D. (1989). Aerosol penetra-tion measurements through protective clothing in small scale simulation tests. Lawrence Livermore National Laboratory. Presented at: 3rd International Symposium on Protection against Chemical Warfare Agents, UMEA, Sweden, June 11–16.

Borm, P. J. A., Schins, R. P. F. & Albrecht, C. (2004). Inhaled particles and lung cancer, Part B: Paradigms and risk assessment. *Int J Cancer, 110*, 3–14.

Bowler, R. M., Roels, H. A., Nakagawa, S., Drezgic, M., Diamond, E., Park, R., Koller, W., Bowler, R. P., Mergler, D., Bouchard, M., Smith, D., Gwiazdam R., Doty, R.L. (2007). Dose-effect relationships between manganese exposure and neurological, neuropsychological and pulmonary function in confined space bridge welders. *Occup Environ Med, 64(3)*, 167–177.

British Standards Institute. (2008). Nanotechnologies, Part 2: Guide to safe handling and disposal of manufactured nanomaterials. PD 6699–2:2007 (www.bsi-global.com/upload/Standards%20&%20 Publications/Nanotechnologies/PD6699-2.pdf).

Brouwer, D. H., Gijsbers, J. H. J. & Lurvink, M. W. M. (2004). Personal exposure to ultrafine particles in the workplace: exploring sampling techniques and strategies. *Ann Occup Hyg, 48(5)*, 439–453.

Brown, D. M., Wilson, M. R., MacNee, W., Stone, V. & Donaldson, K. (2001). Size-dependent proinflammatory effects of ultrafine polystyrene particles: a role for surface area and oxidative stress in the enhanced activity of ultrafines. *Toxicol App Pharmacol, 175(3)*, 191–199.

Brown, J. S., Zeman, K. L. & Bennett, W. D. (2002). Ultra- fine particle deposition and clearance in the healthy and obstructed lung. *Am J Respir Crit Care Med, 166*, 1240–1247.

Brunauer, S., Emmett, P. H. & Teller, E. (1938). Adsorption of gases in multimolecular layers. *J Am Chem Soc, 60*, 309.

Burton, J. (1997). *General methods for the control of airborne hazards*. In: DiNardi SR, ed. The occupational environment—its evaluation and control. Fairfax, VA: American Industrial Hygiene Association.

Castranova, V. (2000). From coal mine dust to quartz: mechanisms of pulmonary pathogenicity. *Inhal Tox, 12*(Suppl 3), 7–14.

Daigle, C. C., Chalupa, D. C., Gibb FRMorrow, P. E., Oberdorster, G., Utell, M. J. & Frampton, M. W. (2003). Ultrafine particle deposition in humans during rest and exercise. *Inhal Toxicol, 15(6)*, 539–552.

De Lorenzo, A. J. D. (1970). *The olfactory neuron and the blood-brain barrier*. In: Wolstenholme GEW, Knight J, eds. Taste and smell in vertebrates. CIBA Foundation Symposium series. London: J&A Churchill, pp. 151–176.

Dockery, D. W, Pope, C. A., Xu, X., Spengler, J. D., Ware, J. H., Fay, M. E., Ferris, B. G. & Speizer, B. E. (1993). An association between air pollution and mortality in six U.S. cities. *N Engl J Med, 329(24)*, 1753–1759.

Donaldson, K. & Stone, V. (2003). Current hypotheses on the mechanisms of toxicity of ultrafine particles. *Ann 1st Super Sanita, 39(3)*, 405–410.

Donaldson, K., Aitken, R., Tran, L., Stone, V., Duffin, R., Forrest, G. & Alexander, A. (2006). Carbon nanotubes: a review of their properties in relation to pulmonary toxicology and workplace safety. *Toxicol Sci,92(1)*, 5–22.

Donaldson, K., Li, X. Y. & MacNee, W. (1998). Ultrafine (nanometer) particle mediated lung injury. *J Aerosol Sci,29(5–6)*, 553–560.

Donaldson, K., Tran, C. L. (2002). Inflammation caused by particles and fibers. *Inhal Toxicol, 14(1)*, 5–27.

Donaldson, K., Tran, L., Jimenez, L. A., Duffin, R., Newby, D. E., Mills, N., MacNee, W. & Stone, V. (2005). Combustion-derived nanoparticles: a review of their toxicology following inhalation exposure.*Part Fibre Toxicol, 2(10)*, 14.

Driscoll, K. E. (1996). *Role of inflammation in the development of rat lung tumors in response to chronic particle exposure*. In: Mauderly JL, McCunney RJ, eds. Particle overload in the rat lung and lung cancer: implications for human risk assessment. Philadelphia, PA: Taylor & Francis, pp. 139–152.

Duffin, R., Tran, C. L., Clouter, A., Brown, D. M., MacNee, W., Stone, V. & Donaldson, K. (2002). The importance of surface area and specific reactivity in the acute pulmonary inflammatory response to particles. *Ann Occup Hyg, 46*, 242–245.

Duffin, R., Tran, L., Brown, D., Stone, V. & Donaldson, K. (2007). Proinflammogenic effects of low-toxicity and metal nanoparticles in vivo and in vitro: highlighting the role of particle surface area and surface reactivity. *Inhal Toxicol,19(10)*, 849–856.

Edlich, R. F., Suber, F., Neal, J. G., Jackson, E. M. & Williams, F. M. (1999). Integrity of powder-free gloves to bacteriophage penetration. *J Biomed Mater Res B Appl Biomater,48*, 755–758.

Elder, A., Gelein, R., Silva, V., Feikert, T., Opanashuk, L., Carter, J., Potter, R., Maynard, A., Ito, Y., Finkelstein, J. & Oberdörster, G. (2006). Translocation of inhaled ultra- fine manganese oxide particles to the central nervous system. *Environ Health Perspect,114*, 1172–1178.

Eninger, R. M., Honda, T., Adhikari, A., Heinonen-Tanski, H., Reponen, T. & Grinshpun, S. A. (2008b). Filter performance of N99 and N95 facepiece respirators against viruses and ultrafine particles. *Ann Occup Hyg,52(5)*, 385–396.

Eninger, R. M., Honda, T., Reponen, T., McKay, R. & Grinshpun, S. A. (2008a) . What does respirator certification tell us about filtration of ultrafine particles? *J Occup Environ Hyg,5*, 286–295.

European Committee for Standardization. (2002).*CEN 943–1:2002: Protective clothing against liquid and gaseous chemicals, aerosols and solid particles*. Performance requirements for ventilated and non- ventilated "gas-tight" (Type 1) and "non-gas-tight" (Type 2) chemical protective suits. Brussels: European Committee for Standardization.

Evans, D. E., Heitbrink, W. A., Slavin, T. J. & Peters, T. M. (2008). Ultrafine and respirable particles in an automotive grey iron foundry. *Ann Occup Hyg,52(1)*, 9–21.

Frampton, M. W., Stewart, J. C., Oberdorster, G., Morrow, P. E., Chalupa, D., Pietropaoli, A. P., Frasier, L. M., Speers, D. M., Cox, C., Huang, L. S. & Utell, M. J. (2006). Inhalation of ultrafine particles alters blood leukocyte expression of adhesion molecules in humans. *Environ Health Perspect,114(1)*, 51–58.

Fuchs NA. (1964). The mechanics of aerosols. Oxford, England: Pergamon Press.

Gao, P., King, W. P. & Shaffer, R. (2007). Review of chamber design requirements for testing of personal protective clothing ensembles. *J Occup & Environ Hyg, 4(8)*, 562–571.

Gardiner, K., van Tongeren, M., Harrington, M. (2001). Respiratory health effects from exposure to carbon black: results of the phase 2 and 3 cross sectional studies in the European carbon black manufacturing industry. *Occup Environ Med, 58(8)*, 496–503.

Gardner, P., Hofacre, K. & Richardson, A. (2004). Comparison of simulated respirator fit factors using aerosol and vapor challenges. *J Occup and Environ Hyg, 1(1)*, 29–38.

Garshick, E., Laden, F., Hart, J. E., Rosner, B., Smith, T. J., Dockery, D. W. & Speizer. F. E. (2004). Lung cancer in railroad workers exposed to diesel exhaust. Environ Health Perspect, *112(15)*, 1539–1543.

Garshick, E., Laden, F., Hart, J. E., Smith, T. J. & Rosner, B. (2006). Smoking imputation and lung cancer in railroad workers exposed to diesel exhaust. *Am J Ind Med,49(9)*, 709–718.

Geiser, M., Rothen-Rutishauser, B., Kapp, N., Schurch, S., Kreyling, W., Schulz, H., Semmler, M., Im Hof, V., Hey- der, J. & Gehr, P. (2005). Ultrafine particles cross cellular membranes by nonphagocytic mechanisms in lungs and in cultured cells. *Environ Health Perspect,113(11)*,1555–1 560.

Golanski, L., Guiot, A. & Tardif, F. (2008). Are conventional protective devices such as fibrous filter media, respirator cartridges, protective clothing and gloves also efficient for nanoaerosols? Brussels, Germany: European Community, Nanotech2 Project.

Goldstein, M., Weiss, H., Wade, K., Penek, J., Andrews, L. & Brandt-Rauf, P. (1987). An outbreak of fume fever in an electronics instrument testing laboratory. *J Occup Med,29*, 746–749.

Granier, J. J. & Pantoya, M. L. (2004). Laser ignition of nano- composite thermites. *Combust Flame, 138*, 373–382.

Han, J. H., Lee, E. J., Lee, J. H., So, K. P., Lee, Y. H., Bae, G. N., Lee, S. B., Ji, J. H., Cho, M. H. & Yu, I. J. (2008). Monitoring multiwalled carbon nanotube exposure in carbon nano- tube research facility. *Inhal Toxicol, 20(8)*, 741–749.

Hart, J. E., Laden, F., Schenker, M. B. & Garshick, E. (2006). Chronic obstructive pulmonary disease mortality in diesel-exposed railroad workers. *Environ Health Perspect, 114(7)*, 1013–1017.

HEI. (2000). Reanalysis of the Harvard Six Cities Study and the American Cancer Society Study of Particulate Air Pollution and Mortality. Cambridge, MA: Health Effects Institute.

Heim, M., Mullins, B., Wild, M., Meyer, J. & Kasper, G. (2005). Filtration efficiency of aerosol particles below 20 nanometers. *Aerosol Sci Tech,39*, 782–789.

Heinrich, U., Fuhst, R., Rittinghausen, S., Creutzenberg, O., Bellmann, B., Koch, W. & Levsen, K. (1995). Chronic inhalation exposure of wistar rats and 2 different strains of mice to diesel-engine exhaust, carbon-black, and titanium-dioxide. *Inhal Toxicol, 7(4)*, 533–5 56.

Heitbrink, W. A., Evans, D. E., Ku, B. K., Maynard, A. D., Slavin, T. J. & Peters, T. M. (forthcoming). Relationships between particle number, surface area, and respirable mass concentration in automotive engine manufacturing. J Occup Environ Hyg.

Heitbrink, W. A., Evans, D. E., Peters, T. M. & Slavin, T. J. (2007). The characteristion and mapping of very fine particles in an engine machining and assembly facility. *J Occup Environ Hyg, 4*, 341–35 1.

Hinds, W. C. (1999). Aerosol technology: properties, behavior, and measurement of airborne particles. 2nd ed. New York: Wiley-Interscience.

Hofacre, K. C. (2006). Aerosol penetration of fabric swatches. Unpublished paper presented at the Elevated Wind Aerosol Conference, Arlington, VA, September 2006..

Hoshino, A., Fujioka, K., Oku, T., Suga, M., Ssaki, Y. & Ohta, T. (2004). Physicochemical properties and cellular toxicity of nanocrystal quantum dots depend on their surface modification. *Nano Lett, 4(11)*, 2163–2169.

HSE. (2004). Horizon scannon information sheet on nanotechnology. Sudbury, Suffolk, United Kingdom: Health and Safety Executive (www.hse.gov/pubns/ hsin1 .pdf).

Huang, S. H., Chen, C. W., Chang, C. P., Lai, C. Y. & Chen, C. C. (2007). Penetration of 4.5 nm to 10 μm aerosol particles through fibrous filters. *J Aero Sci,38*, 719–727.

Hunter, D. D. & Dey, R. D. (1998). Identification and neuropeptide content of trigeminal neurons innervating the rat nasal epithelium. *Neuroscience,83(2)*, 591–599.

Ibald-Mulli, A., Wichmann, H. E., Kreyling, W. & Peters, A. (2002). Epidemiological evidence on health effects of ultrafine particles. *J Aerosol Med Depos,15(2)*, 189–201.

ICRP. (1994). *Human respiratory tract model for radiological protection*. Oxford, England: Pergamon, Elsevier Science Ltd., International Commission on Radiological Protection, Publication No. 66.

International Council on Nanotechnology (ICON). (2006). A survey of current practices in the nanotechnology workplace. (http://cohesion.rice.edu/Centers- AndInst/ICON/ emplibrary/ICONNanotechSurvey- FullReduced.pdf).

ISO. (1995). *Air quality; particle size fraction definitions for health-related sampling, ISO 7708*. Geneva, Switzerland: International Organization for Standardization.

ISO. (2004a). *Protective clothing for use against solid particulates*. Part 2: Test method of determination of inward leakage of aerosols of fine particles into suits. ISO Standard 13982–2. Geneva, Switzerland: International Organization for Standardization.

ISO. (2004b). ISO 16604:2004*: Clothing for protection against contact with blood and body fluids*: determination of resistance of protective clothing materials to penetration by blood-borne pathogens; test method using Phi-X 174 bacteriophage. Geneva, Switzerland: International Organization for Standardization.

ISO. (2006). *Workplace atmospheres; ultrafine, nano- particle and nano-structured aerosols; exposure characterization and assessment*. Geneva, Switzerland: International Standards Organization. Document no. ISO/TC 146/SC 2/WG1 N324, 32 pages.

Jaques. P.A. & Kim, C. S. (2000). Measurement of total lung deposition of inhaled ultrafine particles in healthy men and women. *Inhal Toxicol,12(8)*, 715–731.

Jiang, J., Oberdorster, G., Elder, A., Gelein, R., Mercer, P. & Biswas, P. (2008). Does nanoparticle activity depend upon size and crystal phase? *Nanotox,2(1)*, 33–42.

Johnston, C. J., Finkelstein, J. N., Mercer, P., Corson, N., Gelein, R. & Oberdorster, G. (2000). Pulmonary effects induced by ultrafine PTFE particles. *Toxicol Appl Pharmacol,168*, 208–215.

Keskinen, J., Pietarinen, K. & Lehtimaki, M. (1992). Electrical low pressure impactor. *J Aerosol Sci, 23*, 353–360.

Kim, C. S. & Jaques, P. A. (2004). Analysis of total respiratory deposition of inhaled ultrafine particles in adult subjects at various breathing patterns. *Aerosol Sci Technol,38*, 525–540.

Kim, C. S., Bao, L., Okuyama, K., Shimada, S. & Niinuma, H. (2006). Filtration efficiency of a fibrous filter for nanoparticles. *J Nanopart Res,8*, 215 –221.

Kim, S. E., Harrington, M. S. & Pui, D. Y. H. (2007). Experimental study of nanoparticles penetration through commercial filter media. *J Nanopart Res,9*, 117–125.

Kreiss, K., Mroz, M. M., Zhen, B., Wiedemann, H. & Barna, B. (1997). Risks of beryllium disease related to work processes at a metal, alloy, and oxide production plant. *Occup Environ Med, 54(8)*, 605–612.

Kreyling, W. G., Semmler, M., Erbe, F., Mayer, P., Takenaka, S., Schulz, H., Oberdorster, G. & Ziesenis, A. (2002). Trans-location of ultrafine insoluble iridium particles from lung epithelium to extrapulmonary organs is size dependent but very low. *J Toxicol Environ Health,65(20)*, 1513–1530.

Kreyling, W. G., Semmler-Behnke, M. & Moller, W. (2006). Ultrafine particle-lung interactions: does size matter? *J Aerosol Med,19(1)*,74–83.

Ku, B. K. & Maynard, A. D. (2005). Comparing aerosol surface-area measurements of monodisperse ultrafine silver agglomerates by mobility analysis, transmission electron microscopy and diffusion charging. *J Aerosol Sci,36*, 1108–1124.

Ku, B. K. & Maynard, A. D. (2006). Generation and investigation of airborne silver nanoparticles with specific size and morphology by homogeneous nucleation, coagulation and sintering. *J Aerosol Sci, 37*, 452–470.

Kuhlbusch, T. A. J., Neumann, S. & Fissan, H. (2004). Number size distribution, mass concentration, and particle composition of PM1, PM2.5, and PM10 in bag filling areas of carbon black production. *J Occup Environ Hyg, 1(10)*, 660–671.

Lam, C. W., James, J. T., McCluskey, R. & Hunter, R. L. (2004). Pulmonary toxicity of single-wall carbon nanotubes in mice 7 and 90 days after intratracheal instillation. *Toxicol Sci,77*, 126–134.

Lam, C. W., James, J. T., McCluskey, R., Arepalli, S. & Hunter, R. L. (2006). A review of carbon nanotube toxicity and assessment of potential occupational and environmental health risks. *Crit Rev Toxicol,36*, 189–217.

Lee, C. H., Guo, Y. L, Tsai, P. J., Chang, H. Y., Chen, C. R., Chen, C. W. & Hsiue, T. R. (1997). Fatal acute pulmonary oedema after inhalation of fumes from polytetrafluoro-ethylene (PTFE). *Eur Res J,10*, 1408–1411.

Lee, K. P., Trochimowicz, H. J., Reinhardt, C. F. (1985). Pulmonary response of rats exposed to titanium dioxide (TiO_2) by inhalation for two years. *Toxicol Appl Pharmacol,79*, 179–192.

Lee, S. A., Grinshpun, S. A. & Reponen, T. (2008). Respiratory performance offered by N95 respirators and surgical masks: Human subject evaluation with NaCl aerosol representing bacterial and viral particle size range. *Ann Occup Hyg, 52(3)*, 177–85. Epub 2008 Mar 7.

Li, N., Sioutas, C., Cho, A., Schmitz, D., Misra, C., Sempf, J., Wang, M. Y., Oberley, T., Froines, J. & Nel, A. (2003). Ultrafine particulate pollutants induce oxidative stress and mitochondrial damage. *Environ Health Perspect,111(4)*, 455–460.

Li, Z., Hulderman, T., Salmen, R., Chapman, R., Leonard, S. S., Young, S. H., Shvedova, A., Luster, M. I. & Simeonova, P. P. (2007). Cardiovascular effects of pulmonary exposure to single-wall carbon nanotubes. *Environ Health Perspect,115*, 377–382.

Lippmann, M. (1977). Regional deposition of particles in the human respiratory tract. In: Lee DHK, Murphy S, eds. Handbook of physiology; section IV, environmental physiology. Philadelphia, PA: Williams and Wilkins, pp. 213–232.

Lison, D., Lardot, C., Huaux, F., Zanetti, G. & Fubini, B. (1997). Influence of particle surface area on the toxicity of insoluble manganese dioxide dusts. *Arch Toxicol, 71 (12)*, 725–729.

Long, T. C., Saleh, N., Tilton, R. D., Lowry, G. V. & Veronesi, B. (2006). Titanium dioxide (P25) produces reactive oxygen species in immortalized brain microglia (BV2): implications for nanoparticle neurotoxicity. *Environ Sci & Technol, 40(14)*, 4346–4352. Epub 16 June 2006.

Lovric, J., Bazzi, H. S., Cuie, Y., fortin, G. R. A., Winnik, F. M. & Maysinger, D. (2005). Differences in subcellular distribution and toxicity of green and red emitting CdTe quantum dots. *J Mol Med,83(5)*, 377–385. Epub 2005 Feb 2.

Lux Research. (2007). The Nanotech Report, 5th Edition. New York: Lux Research.

Marple, V. A., Olson, B. A. & Rubow, K. L. (2001). Inertial, gravitational, centrifugal, and thermal collection techniques. In: Baron PA, Willeke K, eds. Aerosol measurement: principles, techniques and applications. New York: John Wiley and Sons, pp. 229–260.

Martin, S. & Moyer, E. (2000). Electrostatic respirator filter media: filter efficiency and most penetrating particle size effects. *App Occup Environ Hyg,15(8)*, 609–617.

Maynard, A. D. (2003). Estimating aerosol surface area from number and mass concentration measurements. *Ann Occup Hyg,47(2)*, 123–144.

Maynard, A. D., Baron, P. A., Foley, M., Shvedova, A. A., Kisin, E. R. & Castranova, V. (2004). Exposure to carbon nano- tube material: aerosol release during the handling of unrefined single walled carbon nanotube material. *J Toxicol Environ Health, 67(1)*, 87–107.

Maynard, A. M. & Kuempel, E. D. (2005). Airborne nano- structured particles and occupational health. *J Nano- particle Research,7(6)*, 587–614.

Mercer, R. R., Scabilloni, J., Wang, L., Kisin, E., Murray, A. R., Schwegler-Berry, D., Shvedova, A. A. & Castranova, V. (2008). Alteration of depositon pattern and pulmonary response as a result of improved dispersion of aspirated single walled carbon nanotubes in a mouse model. *Am J Physiol Lung Cell Mol Physiol,294*, L87–L97.

Methner, M. M. (2008). Engineering case reports: Old L, ed. Effectiveness of local exhaust ventilation (LEV) in controlling engineered nanomaterial emissions during reactor cleanout operations. *J Occup Environ Hyg,5(6)*, D63–D69.

Methner, M. M., Birch, M. E., Evans, D. E., Ku, B. K., Crouch, K. G. & Hoover, M. D. (2007). Mazzulxeli LF, ed: Case study: Identification and characterization of potential sources of worker exposure to carbon nanofibers during polymer composite laboratory operations. *J Occup Environ Hyg,4(12)*, D125–D130.

Misra, C., Singh, M., Shen, S., Sioutas, C. & Hall, P. M. (2002). Development and evaluation of a personal cascade impactor sampler (PCIS). *J Aerosol Sci,33(7)*, 1027–1048.

Moller, W., Brown, D. M., Kreyling, W. G. & Stone, V. (2005). Ultrafine particles cause cytoskeletal dysfunctions in macrophages: role of intracellular calcium.*Part Fibre Toxicol,2(7)*, 12.

Moller, W., Hofer, T., Ziesenis, A., Karg, E. & Heyder, J. (2002). Ultrafine particles cause cytoskeletal dysfunctions in macrophages. *Toxicol Appl Pharmacol,182(3)*, 197–207.

Monteiro-Riviere, N. A., Nemanich, R. J., Inman, A. O., Wang, Y. Y. & Riviere, J. E. (2005). Multi-walled carbon nanotube interactions with human epidermal keratinocytes. *Toxicol Lett, 155(3)*, 377–384.

Moshammer, H. & Neuberger, M. (2003). The active surface of suspended particles as a predictor of lung function and pulmonary symptoms in Austrian school children. *Atmos Environ,37(13)*, 1737–1744.

Mossman, B. & Churg, A. (1998). Mechanisms in the pathogenesis of asbestosis and silicosis. *Am J Respir Crit Care Med, 157*, 1666–1680.

Muller, J., Huaux, F., Moreau, N., Misson, P., Heilier, J. F., Delos, M., Arras, M., Fonseca, A., Nagy, J. B. & Lison, D. (2005). Respiratory toxicity of multi-wall carbon nanotubes. *Toxicol Appl Pharmacol,207*, 221–231.

Murray, A. R., Kisin, E., Kommineni, C., Kagan, V. E., Castranova, V. & Shvedova, A. A. (2007). Single-walled carbon nanotubes induce oxidative stress and infl ammation in skin. *Toxicologist,96*, A1406.

National Fire Protection Association (NFPA). (2007). NFPA 1994 standard on protective ensembles for first responders to CBRN terrorism incidents.

Nel, A., Xia, T., Madlen, L. & Li, W. (2006). Toxic potential of materials at the nanolevel. *Science,311,* 622–627.

Nemmar, A., Hoet, P. H. M., Vanquickenborne, B., Dins-dale, D., Thomeer, M., Hoylaerts, M. F., Vanbilloen, H., Mortelmans, L. & Nemery, B. (2002). Passage of inhaled particles into the blood circulation in humans. *Circulation,105*, 411–414.

NIOSH. (1990). NIOSH testimony on the Occupational Safety and Health Administration proposed rule on health standards: methods of compliance. Cincinnati, OH: U.S. Department of Health and Human Services, Centers for Disease Control, National Institute for Occupational Safety and Health.

NIOSH. (1994). NIOSH manual of analytical methods (NMAM®). 4th ed. By Schlecht PC, O'Connor PF, eds. Cincinnati, OH: U.S. Department of Health and Human Services, Centers for Disease Control and Prevention, National Institute for Occupational Safety and Health, DHHS (NIOSH) Publication 94–113 (www.cdc.gov/niosh/nmam/).

NIOSH. (1996). NIOSH guide to the selection and use of particulate respirators certified under 42 CFR 84. Cincinnati, OH: U.S. Department of Health and Human Services, Centers for Disease Control and Prevention, National Institute for Occupational Safety and Health, DHHS (NIOSH) Publication No. 96–101.

NIOSH. (2003). Filtration and air-cleaning systems to protect building environments. Cincinnati, OH: U.S. Department of Health and Human Services, Centers for Disease Control and Prevention, National Institute for Occupational Safety and Health, DHHS (NIOSH) Publication No. 2003–136.

NIOSH. (2004). *NIOSH respirator selection logic*. Cincinnati, OH: U.S. Department of Health and Human Services, Centers for Disease Control and Prevention, National Institute for Occupational Safety and Health, DHHS (NIOSH) Publication No. 2005–100 (www.cdc.gov/niosh/docs/2005-100/).

NIOSH. (2005a) . Procedure No. RCT–APR–STP–0057, 0058, 0059, Revision 1.1. Cincinnati, OH: Department of Health and Human Services, Centers for Disease Control and Prevention, National Institute for Occupational Safety and Health (www.cdc.gov/niosh/ npptl/stps/pdfs/RCT-APR-0057%2058%2059.pdf).

NIOSH. (2005b). Procedure No. RCT–APR– STP–0051, 0052, 0053, 0054, 0055, 0056, Revision 1.1. Cincinnati, OH: Department of Health and Human Services, Centers for Disease Control and Prevention, National Institute for Occupational Safety and Health (www.cdc.gov/niosh/npptl/stps/pdfs/ RCT-APR-005 1%2052%2053%2054%2055% 2056. pdf).

NIOSH. (2007). Docket Number 036: Total Inward Leakage Test for Half-Mask Air-Purifying Particulate Respirators; Total Inward Leakage Test for Half-mask Air-purifying Particulate Respirators. Procedure No.

NIOSH. (2009). *Current intelligence bulletin: interim guidance on medical screening of workers potentially exposed to engineered nanoparticles*. Cincinnati, OH: U.S. Department of Health and Human Services, Centers for Disease Control and Prevention, National Institute for Occupational Safety and Health (www.cdc.gov/niosh/docs/2009-116/).

Oberdörster, G., Ferin, J. & Lehnert, B. E. (1994a). Correlation between particle-size, in-vivo particle persistence, and lung injury. *Environ Health Perspect,102(S5)*, 173–179.

Oberdörster, G., Ferin, J., Gelein, R., Soderholm, S. C. & Finkelstein, J. (1992). Role of the alveolar macrophage in lung injury—studies with ultrafine particles. *Environ Health Perspect*, *97*, 193–199.

Oberdörster, G., Ferin, J., Soderholm, S., Gelein, R., Cox, C., Baggs, R. & Morrow, P. E. (1994b). Increased pulmonary toxicity of inhaled ultrafine particles: due to lung overload alone? *Ann Occup Hyg,38*(Suppl 1), 295–302.

Oberdörster, G., Gelein, R. M., Ferin, J. & Weiss, B. (1995). Association of particulate air pollution and acute mortality: involvement of ultrafine particles? *Inhal Toxicol, 7(1)*, 111–124.

Oberdörster, G., Maynard, A., Donaldson, K., Castranova, V., Fitzpatrick, J., Ausman, K., Carter, J., Karn, B., Kreyling, W., Lai, D., Olin, S., Monteiro-Riviere, N., Warheit, D. & Yang, H. (2005b). Principles for characterizing the potential human health effects from exposure to nanomaterials: elements of a screening strategy. *Part Fibre Toxicol,2*, 8.

Oberdörster, G., Oberdörster, E. & Oberdörster, J. (2005a). Nanotoxicology: an emerging discipline evolving from studies of ultrafine particles. *Environ Health Perspect,113(7)*, 823–839.

Oberdörster, G., Sharp, Z., Atudorei, V., Elder, A., Gelein, R., Lunts, A., Kreyling, W. & Cox, C. (2002). Extrapulmonary translocation of ultrafine carbon particles following whole-body inhalation exposure of rats. *J Toxicol Environ Health 65 Part A, (20)*, 1531–1543.

Oberdörster, G., Sharp, Z., Atudorei, V., Elder, A., Gelein, R., Kreyling, W. & Cox, C. (2004). Translocation of inhaled ultrafine particles to the brain. *Inhal Toxicol, 16(6–7)*, 437–445.

Oberdörster, G., Yu (1990). The carcinogenic potential of inhaled diesel exhaust: a particle effect? *J Aerosol Sci,21*, S397–S401.

Park, R. M., Bowler, R. M., Eggerth, D. E., Diamond, E., Spencer, K. J., Smith, D. & Gwiazda, R. (2006). Issues in neurological risk assessment for occupational exposures: the Bay Bridge welders. *Neurotoxicology,27(3)*, 373–384.

Penttinen, P., Timonen, K. L., Tiittanen, P., Mirme, A., Russkanen, J. & Pekkanen, J. (2001). Ultrafine particles in urban air and respiratory health among adult asthmatics. *Eur Respir J,17(3)*, 428–435.

Peters, A., Dockery, D. W., Heinrich, J. & Wichmann, H. E. (1997). Short-term effects of particulate air pollution on respiratory morbidity in asthmatic children. *Eur Respir J,10(4)*, 872–879.

Peters, A., von Klot, S., Heier, M., Trentinaglia, I., Hormann, A., Wichmann, H. E. 7 Lowel, H. (2004). Exposure to traffic and the onset of myocardial infarction. *N Engl J Med,351(17)*, 1721–1730.

Peters, T. M., Heitbrink, W. A., Evans, D. E., Slavin, T. J. 7 Maynard, A. D. (2006). The mapping of fine and ultrafine particle concentrations in an engine machining and assembly facility. *Ann Occup Hyg,50(3)*, 249–257.

Plog, B. A., Niland, J. & Quinlan, P. J., eds. (2002). Fundamentals of industrial hygiene, 5th ed. Itasca, IL: National Safety Council.

Poland, C. A., Duffin, R., Kinloch, I., Maynard, A., Wallace, W. A. H., Seaton, A., Stone, V., Brown, S., MacNee, W. & Donaldson, K. (2008). Carbon nanotubes introduced into the abdominal cavity of mice show asbestos- like pathogenicity in a pilot study. *Nat Nanotech,3(7)*, 423–428. Epub 2008 May 20. (www.nature.com/ naturenanotechnology).

Pope, C. A., Burnett, R. T., Thun, M. J., Calle, E. E., Krewski, E., Ito, K. & Thurston, G. D. (2002). Lung cancer, cardiopulmonary mortality and long term exposure to fine particulate air pollution. *JAMA,287(9)*, 1 132–1141.

Pope, C. A., Burnett, R. T., Thurston, G. D., Thun, M. J., Calle, E. E., Krewski, D. & Godleski, J. J. (2004). Cardiovascular mortality and long-term exposure to particulate air pollution: epidemiological evidence of general pathophysiological pathways of disease. *Circulation,109(1)*, 71–74.

Porter, A. E., Gass, M., Muller, K., Skepper, J. N., Midgley, P. & Welland, M. (2007a). Direct imaging of single-walled carbon nanotubes in cells. *Nat Nanotech*, *2*, 713–717.

Porter, A. E., Gass, M., Muller, K., Skepper, J. N., Midgley, P. & Welland, M. (2007b). Visualizing the uptake of C60 to the cytoplasm and nucleus of human monocyte-derived macrophage cells using energy-filtered transmission electron microscopy and electron to-mography. *Environ Sci Technol*, *41(8)*, 3012–3017.

Pott, F. & Roller, M. (2005). Carcinogenicity study with nineteen granular dusts in rats. *Eur J Oncol,10(4)*, 249–28 1.

Pritchard, D. K. (2004). Literature review—explosion hazards associated with nanopowders. United Kingdom: Health and Safety Laboratory, HSL/2004/12.

Pui, D. Y. H. & Kim, S. C. (2006). *Penetration of nanoparticles through respirator filter media*. Minneapolis, MN: University of Minnesota, Mechanical Engineering Department, Particle Technology Laboratory. NIOSH Contract No. 254–2005–M–1 1698 for National Personal Protective Technology Division.

Ratherman, S. (1996). Methods of control. In: Plog B, ed. Fundamentals of industrial hygiene. Itasca, IL: National Safety Council.

RCT–APR–STP–0068, January 31, 2007 (www.cdc. gov/niosh/docket/NIOSHdo cket0036.html).

Rengasamy, S., Eimer, B. C. & Shaffer, R. E. (2008b). Nano- particle filtration performance of commercially available dust masks. *J Int Soc Res Prot,25*, 27.

Rengasamy, S., King, W. P., Eimer, B. & Shaffer, R. E. (2008a). Filtration performance of NIOSH-approved N95 and P100 filtering-facepiece respirators against 4–30 nanometer size nanoparticles. *J Occup Environ Hyg,5(9)*, 556–564.

Rengasamy, S., Verbofsky, R., King, W. P. & Shaffer, R. E. (2007). Nanoparticle penetration through NIOSHapproved N95 filtering-facepiece respirators. *J Int Soc Res Prot,24*, 49–59.

Renwick, L. C., Brown, D., Clouter, A. & Donaldson, K. (2004). Increased inflammation and altered macrophage chemotactic responses caused by two ultra- fine particles. *Occup Environ Med,61*, 442–447.

Richardson, A. W., Eshbaugh, J. P., Hofacre, K. C. & Gardner, P. D. (2005). Respirator filter efficiency testing against particulate and biological aerosols under moderate to high flow rates. Columbus, OH: Battelle Memorial Institue. Contract Report No. SP0700–00–D–3 180.

Ruckerl, R., Ibald-Mulli, A., Koenig, W., Schneider, A., Woelke, G., Cyrys, J., Heinrich, J., Marder, V., Frampton, M., Wichmann, H. E. & Peters, A. (2006). Air pollution and markers of inflammation and coagulation in patients with coronary heart disease. *Am J Respir Crit Care Med,173(4)*, 432–441.

Ryman-Rasmussen, J. P., Riviere, J. E. & MonteiroRiviere, J. E. (2006). Penetration of intact skin by quantum dots with diverse physicochemical properties. *Toxicol Sci,91(1)*, 159–165.

Sager, T., Porter, D. & Castranova, V. (2008). Pulmonary response to intratracheal instillation of fine or ultra- fine carbon black or titanium dioxide: Role of surface area. *Toxicologist,102*, A1491.

Sayes, C. M., Fortner, J. D., Guo, W., Lyon, D., Boyd, A. M., Ausman, K. D., Tao, Y. J., Sitharaman, B., Wilson, L. J., Hughes, J. B., West, J. L. & Colvin, V. L. (2004). The differential cytotoxicity of water-soluble fullerenes. *Nano Letters,4(10)*, 1881–1887.

Sayes, C. M., Liang, F., Hudson, J. L., Mendez, J., Guo, W., Beach, J. M., Moore, V. C., Doyle, C. D., West, J. L., Billups, W. E., Ausman, K. D. & Colvin, V. L. (2005). Functionalization density dependence of single-walled carbon nanotubes cytotoxicity in vitro. *Toxicol Lett,161(2)*, 135–142.

Sayes, C. M., Wahi, R., Kurian, P. A., Liu, Y., West, J. L., Ausman, K. D., Warheit, D. B. & Colvin, V. L. (2006). Correlating nanoscale titania structure with toxicity: a cytotoxicity and inflammatory response study with human dermal fibroblasts and human lung epithelial cells. *Toxicol Sci,92(1)*, 174–185.

Schneider, T., Cherrie, J. W., Vermeulen, R. & Kromhout, H. (1999). Conceptual model for assessment of dermal exposure. *J Occup Environ Med, 56*, 765–773.

Schneider, T., Cherrie, J. W., Vermeulen, R. & Kromhout, H. (2000). Dermal exposure assessment. *Ann Occup Hyg, 44(7)*, 493–499.

Schulte, P. A., Trout, D., Zumwalde, R., Kuempel, E., Geraci, C., Castranova, V., Mundt, D. J., Mundt, K. A. & Halperin, W. E. (2008b). Options for occupational health surveillance of workers potentially exposed to engineered nanoparticles: state of the science. *J Occup Environ Med,50*, 517–526.

Schulte, P., Geraci, C., Zumwalde, R., Hoover, M. & Kuempel, E. (2008a). Occupational risk management of engineered nanoparticles. *J Occup Environ Hyg, 5*, 239–249.

Seinfeld, J. A. & Pandis, S. N. (1998). Atmospheric chemistry and physics. New York: John Wiley and Sons.

Semmler, M., Seitz, J., Erbe, F., Mayer, P., Heyder, J.,Oberdorster, G. & Kreyling, W. G. (2004). Long-term clearance kinetics of inhaled ultrafine insoluble iridium particles from the rat lung, including transient translocation into secondary organs. *Inhal Toxicol,16(6–7)*, 453–459.

Shalev, I., Barker, R. L., McCord, M. G., Tucker, P. A. & Lisk, B. R. (2000). Protective textile particulate penetration screening. Performance of protective clothing: 7th Symposium, ASTM STP 1386, West Conshohocken, PA: *American Society for Testing and Materials, ASTM,* 155–161.

Shiohara, A., Hoshino, A., Hanaki, K., Suzuki, K. & Yamamoto, K. (2004). On the cytotoxicity of quantum dots. *Microbiol Immunol,48(9)*, 669–675.

Shvedova, A. A., Kisin, E. R., Mercer, R., Murray, A. R., Johnson, V. J., Potapovich, A. I. & Tyurina, Y. Y, Gorelik, O., Arepalli, S., Schwegler-Berry, D. (2005). Unusual inflammatory and fibrogenic pulmonary responses to single walled carbon nanotubes in mice. *Am J Physiol Lung Cell Mol Physiol,289(5)*, L698–708. Epub 2005 Jun 10.

Shvedova, A. A., Kisin, E. R., Murray, A. R., Gandelsman, V. Z. & Maynard, A. D., Baron, P. A. & Castranova, V. (2003). Exposure to carbon nanotube material: assessment of the biological effects of nanotube materials using human keratinocyte cells. *J Toxicol Environ Health,66(20)*, 1909–1926.

Shvedova, A. A., Kisin, E. R., Murray, A. R., Schwegler-Berry, D., Gandelsman, V. Z., Baron, P., Maynard, A., Gunther, M. R. & Castranova, V. (2004). Exposure of human bronchial epithelial cells to carbon nanotubes cause oxidative stress and cytotoxicity. In: *Proceedings of the Society for Free Radical Research Meeting*, Paris, France: Society for Free Radical Research International, European Section, June 26–29, 2003.

Shvedova, A. A., Kisin, E., Murray, A. R., Johnson, V., Gorelik, O., Arepalli, S., Hubbs, A. F., Mercer, R. R., Stone, S., Frazer, D., Chen, T., Deye, G., Maybnard, A., Baron, P., Mason, R., Kadiiska, M., Stadler, K., Mouithys-Mickalad, A., Castranova, V., Kaagan, V. E. (2008). Inhalation of carbon nanotubes induces oxidative stress and cytokine response causing respiratory impairment and pulmonary fibrosis in mice. *Toxicologist,102*, A1497.

Shvedova, A. A., Sager, T., Murray, A., Kisin, E., Porter, D. W., Leonard, S. S., Schwegler-Berry, D., Robinson, V. A. & Castranova, V. (2007). Critical issues in the evaluation of possible effects resulting from airborne nanoparticles. In: Monteiro-Riviere N and Tran L (eds). *Nanotechnology: characterization, dosing and health effects*. Philadelphia, PA: Informa Healthcare, pp. 22 1–232.

Sriram, K., Porter, D. W., Tsuruoka, S., Endo, M., Jefferson, A. M., Wolfarth, M. G., Rogers, G. M., Castranova, V. & Luster, M. I. (2007). Neuroinflammatory response following exposure to engineered nanomaterials. *Toxicologist,96*, A1390.

Steenland, K., Deddens, J. & Stayner, L. (1998). Diesel exhaust and lung cancer in the trucking industry: exposure-response analyses and risk assessment. *Am J Ind Med,34(3)*, 220–228.

Stcfaniak, A. B., Hoover, M. D., Dickerson, R. M., Peterson, E. J., Day, G. A., Breysse, P. N., Kent, M. S. & Scripsick, R. C. (2003). Surface area of respirable beryllium metal, oxide, and copper alloy aerosols and implications for assessment of exposure risk of chronic beryllium disease. *Am Ind Hyg Assoc J,64*, 297–305.

Stevens, G. & Moyer, E. (1989). Worst case aerosol testing parameters: I. sodium chloride and dioctyl phthalate aerosol filter efficiency as a function of particle size and flow rate. *Am Ind Hyg Assoc J,50(5)*, 257–264.

Takagi, A., Hirose, A., Nishimura, T., Fukumori, N., Ogata, A., Ohashi, N., Kitajima, S. & Kanno, J. (2008). Induction of mesothelioma in p53+/- mouse by intraperitoneal application of multi-wall carbon nanotube. *J Toxicol Sci,33(1)*, 105–16.

Takenaka, S., Karg, D., Roth, C., Schulz, H., Ziesenis, A., Heinzmann, U., Chramel, P. & Heyder, J. (2001). Pulmonary and systemic distribution of inhaled ultrafine silver particles in rats. *Environ Health Persp*, *109*(suppl. 4), 547–551.

The Royal Society, The Royal Academy of Engineering (2004). Nanoscience and nanotechnologies. London, UK: The Royal Society and The Royal Academy of Engineering (www.nanotec.org.uk/finalReport.htm).

Thomas, K., Aguar, P., Kawasaki, H., Morris, J., Nakanish, J. & Savage, N. (2006). Research strategies for safety evaluation of nanomaterials, part VIII: international efforts to develop risk-based safety evaluations for nanomaterials. *Toxic Sci,92(1)*, 23–32.

Timonen, K. L., Hoek, G., Heinrich, J., Bernard, A., Brunekreef, B., de Hartog, J., Hameri, K., Ibald-Mulli, A., Mirme, A., Peters, A., Tiittanen, P., Kreyling, W. B. & Pekkanen, J. (2004). Daily variation in fine and ultrafine particulate air pollution and urinary concentrations of lung Clara cell protein CC16. *Occup Environ Med,61(11)*, 908–914.

Tinkle, S. S., Antonini, J. M., Rich, B. A., Robert, J. R., Salmen, R., DePree, K. & Adkins, E. J. (2003). Skin as a route of exposure and sensitization in chronic beryllium disease. *Environ Health Perspect,111(9)*, 1202–1208.

Törnqvist, H., Mills, N. L., Gonzalez, M., Miller, M. R., Robinson, S. D., Megson, I. L., Macnee, W., Donaldson, K., Söderberg, S., Newby, D. E., Sandström, T. & Blomberg, A. (2007). Persistent endothelial dysfunction in humans after diesel exhaust inhalation. *Am J Respir Crit Care Med,176(4)*, 395–400.

Tran, C. L., Cullen, R. T., Buchanan, D., Jones, A. D., Miller, B. G., Searl, A., Davis, J. M. G. & Donaldson, K. (1999). Investigation and prediction of pulmonary responses to dust, part II. In: Investigations into the pulmonary effects of low toxicity dusts. Contract Research Report 216/1999. Suffolk, UK: Health and Safety Executive.

Tran, C., Buchanan, L. D., Cullen, R. T., Searl, A., Jones, A. D. & Donaldson, K. (2000). Inhalation of poorly soluble particles. II. Influence of particle surface area on inflammation and clearance. *Inhal Toxicol*, *12(12)*, 1 113–1126.

TSI. (2005). Mechanisms of filtration for high efficiency fibrous filters. Application Note ITI–041, TSI Incorporated (www.tsi.com/AppNotes/appnotes.aspx?Cid=24&Cid2=195&Pid=33&lid=439&file=iti_041#mech).

US DOE. (2007) . Approach to Nanomaterial ES&H, U.S. Department of Energy's Nanoscale Science Research Centers. Washington, DC: U.S. Department of Energy.

VanOsdell, D. W., Liu, B. Y. H., Rubow, K. L., Pui, D. Y. H. (1990). Experimental-study of submicrometer and ultrafine particle penetration and pressure-drop for high-efficiency filters. *Aerosol Sci Technol,12(4),* 91 1–925.

Vaughan, N. P., Milligan, B. D. & Ogden, T. L. (1989). Filter weighing reproducibility and the gravimetric detection limit. *Ann Occup Hyg*, *33(3)*, 331–337.

Wang, L., Castranova, V., Rojanasakul, Y., Lu, Y., Scabilloni, J. & Mercer, R. R. (2008). Direct fibrogenic effects of dispersed single walled carbon nanotubes on human lung fibroblasts. *Toxicologist,102*, A1499.

Wang, Z. M. & Gao, P. (2007). A study on magnetic passive aerosol sampler for measuring aerosol particle penetration through protective ensembles. Unpublished paper presented at the 26th Annual Conference of the American Association for Aerosol Research. Reno, NV, September 24–28.

Warheit, D. B., Laurence, B. R., Reed, K. L., Roach, D. H., Reynolds, G. A. & Webb, T. R. (2004). Comparative pulmonary toxicity assessment of single-wall carbon nanotubes in rats. *Toxicol Sci,77*, 117–125.

Warheit, D. B., Webb, T. R., Reed, K. L., Frerichs, S. & Sayes, C. M. (2007). Pulmonary toxicity study in rats with three forms of ultrafine-TiO_2 particles: differential responses related to surface properties. *Toxicology,230*, 90–104.

Warheit, D. B., Webb, T. R., Sayes, C. M., Colvin, V. L. & Reed, K. L. (2006). Pulmonary instillation studies with nanoscale TiO_2 rods and dots in rats: toxicity is not dependent upon particle size and surface area.*Toxicol Sci,91(1)*, 227–236.

Wiebel, A., Bouchet, R., Boulch, F. & Knauth, P. (2005). The big problem of particle size: a comparison of methods for determination of particle size in nanocrystalline anatase powers. *Chem Mater*, *17*, 2378–2385.

Woo, K. S., Chen, D. R., Pui, D. Y. H. & Wilson, W. E. (2001). Use of continuous measurements of integral aerosol parameters to estimate particle surface area. *Aerosol Sci Technol,34*, 57–65.

Wood, J. P. (2001). Containment in the pharmaceutical industry. New York: Marcel Dekker, Inc.

Sources of Additional Information

Aitken R. J., Creely, K. S. & Tran, C. L. (2004). Nanoparticles: an occupational hygiene review. HSE Research Report 274. UK Health and Safety Executive (www.hse.gov.uk/research/rrhtm/rr274.htm).

Balazy, A., Podgórski, A. & Grado, L. (2004). Filtration of nanosized aerosol particles in fibrous filters: I. Experimental results. Warsaw Poland: Warsaw University of Technology, Department of Chemical and Process Engineering.

Baron, P. A. & Willeke, K. (2001). Aerosol measurement. Principles, techniques and applications. New York: Wiley-Interscience.

Blake, T., Castranova. V., Schwegler-Berry. D., Baron. P., Deye. G. J., Li. C. H. & Jones, W. (1998). Effect of fiber length on glass microfiber cytotoxicity. *J Toxicol Environ Health, 54*(Part A)*(4)*, 243–259.

Borm. P. J. A. (2002). Particle toxicology: from coal mining to nanotechnology. *Inhal Toxicology,14*, 311–324.

Byrd. D. M. & Cothern, C. R. (2000). Introduction to risk analysis. Rockville, MD: Government Institutes.

Castranova, V. (1998). Particles and airways: basic biological mechanisms of pulmonary pathogenicity. *Appl Occup Environ Hyg,13(8)*, 613–616.

Conhaim, R. L., Eaton, A., Staub, N. C. & Heath, T. D. (1988). Equivalent pore estimate for the alveolar-airway barrier in isolated dog lung. *J Appl Physiol,64(3)*, 1 134–1142.

Driscoll, K. E., Costa, D. L., Hatch, G., Henderson, R., Oberdörster, G., Salem, H. & Schlesinger, R. B. (2000). Intratracheal instillation as an exposure technique for the evaluation of respiratory tract toxicity: uses and limitations. *Toxicol Sci,55(1)*, 24–35.

EC (2004). Nanotechnologies: a preliminary risk analysis on the basis of a workshop organized in Brussels, March 1–2, 2004, by the Health and Consumer Protection Directorate General of the European Commission. Brussels, Germany: European Commission (http://www.certh.gr/dat/7CF3F1 C6/file. pdf?633 556363252496250).

EPA (1992). Guidelines for exposure assessment. Washington, DC: U.S. Environmental Protection Agency.

Faux, S. P., Tran, C. L., Miller, B. G., Jones, A. D., Monteiller, C. & Donaldson, K. (2003). In vitro determinants of particulate toxicity: the dose-metric for poorly soluble dusts. Research report 154. Suffolk, UK: Health and Safety Executive.

Ferin, J. & Oberdörster, G. (1992). Translocation of particles from pulmonary alveoli into the interstitium. *J Aerosl Med, 5(3)*, 179–187.

Gibbs, J. P., Crump, K. S., Houck, D. P., Warren, P. A. & Mosley, W. S. (1999). Focused medical surveillance: a search for subclinical movement disorders in a cohort of U.S. workers exposed to low levels of manganese dust. *Neurotoxicol,20(2–3)*, 299–3 13.

Hardman, R. (2006). A toxicologic review of quantum dots: toxicity depends on physicochemical and environmental factors. *Environ Health Perspect,114(2)*, 165–172.

Hayes, A. W., ed. (2001). Principles and methods of toxicology. 4th ed. Philadelphia, PA: Taylor and Francis.

Hoet, P. H. M., Bruske-Hohlfeld, I. & Salata, O. V. (2004). Nanoparticles-known and unknown health risks. *J Nanobiotech,2(12)*, 1–15.

Hofmann, W., Sturm, R., Winkler-Heil, R. & Pawlak, E. (2003). Stochastic model of i ultrafi ne particle deposition and clearance in the human respiratory tract. *Radiat Prot Dosim,105(1–4)*, 77–80.

Hood, E. (2004). Nanotechnology: looking as we leap. Environ Health Perspect 1 12(13):A741–A749.

Huckzko, A. & Lange, H. (2001). Carbon nanotubes: experimental evidence for a null risk of skin irritation and allergy. *Fullerene Sci Tech,9(2)*, 247–250.

Huckzko, A., Lange, H., Calko, E., Grubeck-Jaworska, H. & Droszez, P. (2001). Physiological testing of carbon nanotubes: are they asbestos-like? *Fullerene Sci Tech, 9(2)*, 25 1–254.

Keller, A., Fierz, M., Siegmann, K., Siegmann, H. C. & Filippov, A. (2001). Surface science with nanosized particles in a carrier gas. *J Vac Sci Technol A,19(1)*, 1–8.

Klaassen, C. D., Amdur, M. O. & Doull, J. (1986). Casarett and Doull's toxicology: the basic science of poisons. New York: Macmillan Publishing Company.

Last, J. M. (1983). A dictionary of epidemiology. New York: Oxford University Press.

Lee, K. W. & Liu, B. Y. H. (1981). Experimental study of aerosol filtration by fibrous filters. *Aerosol Sci Technol,1(1)*, 35–46.

Lee, K. W. & Liu, B. Y. H. (1982). Theoretical study of aerosol filtration by fibrous filters. *Aerosol Sci Technol,1(2)*, 147–162.

Li, N., Sioutas, C., Cho, A., Schmitz, D., Misra, C., Sempf, J., Wang, M. Y., Oberley, T., Froines, J. & Nel, A. (2003). Ultrafine particulate pollutants induce oxidative stress and mitochondrial damage. *Environ Health Perspect,111(4)*, 455–460.

Maynard, A. D. (2001). Experimental determination of ultrafine TiO2 de-agglomeration in surrogate pulmonary surfactant—preliminary results. Inhaled Particles IX, Cambridge, UK: The British Occupational Hygiene Society.

Maynard, A. D. (2002). Experimental determination of ultrafine TiO_2 de-agglomeration in surrogate pulmonary surfactant—preliminary results. *Ann Occup Hyg,46*(Suppl 1), 197–202.

NCI, (1979). Bioassay of titanium dioxide for possible carcinogenicity. Technical Report Series No. 97. Bethesda, MD: National Cancer Institute.

Nel, A., Xia, T., Madler, L. & Li, N. (2006). Toxic potential of materials at the nanolevel. *Science, 311*, 622–627.

Nemmar, A., Vanbilloen, H., Hoylaerts, M. F., Hoet, P. H. M., Verbruggen, A. & Nemery, B. S. (2001). Passage of intratracheally instilled ultrafine particles from the lung into the systemic circulation in hamster. *Am J Respir Crit Care Med,164(9)*, 1665–1668.

NFPA. (1994). Standard on protective ensembles for first responders to CBRN terrorism incidents. Quincy, MA: National Fire Protection Association.

Nichols, G., Byard, S., Bloxham, M. J., Botterill, J., Dawson, N. J., Dennis, A., Diart, V., North, N. C. & Sherwood, J. D. (2002). A review of the terms agglomerate and aggregate with a recommendation for nomenclature used in powder and particle characterization. *J Pharm Sci,91(10)*, 2103–2109.

NIOSH. (1987). NIOSH guide to industrial respiratory protection. Cincinnati, OH: U.S. Department of Health and Human Services, Centers for Disease Control and Prevention, National Institute for Occupational Safety and Health, DHHS (NIOSH) Publication No. 87–116, Appendix B.

NRC. (1994). Science and judgement in risk assessment. Washington D.C.: National Academy Press.

Oberdörster, E. (2004). Manufactured nanomaterials (Fullerenes, C60) induce oxidative stress in brain of juvenile largemouth bass. *Environ Health Perspect,112(10)*, 1058–1062.

Oberdörster, G., Ferin, J., Finkelstein, J. & Soderholm, S. (1992). Thermal-degradation events as healthhazards—Particle Vs gas-phase effects, mechanistic studies with particles. *Acta Astronaut,27*, 251–256.

OSHA. (1998). Occupational Safety and Health Administration, OSHA 3143 (revised): industrial hygiene. Washington, DC: U.S. Department of Labor, Occupational Safety and Health Administration.

Patty. (2000). Patty's industrial hygiene and toxicology. New York: John Wiley and Sons.

Pflucker, F., Wendel, V., Hohenberg, H., Gärtner, E., Will, T., Pfeiffer, S., Wept, F. & Gers-Barlag, H. (2001). The human stratum corneum layer: an effective barrier against dermal uptake of different forms of topically applied micronised titanium dioxide. *Skin Pharmacol Appl Skin Physiol, 14*(Suppl 1), 92–97.

Pui, D. Y. H. & Chen, D. (2002). Experimental and modeling studies of nanometer aerosol filtration. Sponsored by the Department of Energy. Minneapolis, MN: University of Minnesota, Particle Technology Laboratory. Grant No. DOEIDE -FG02–98ER1490.

Rao, G. V. S., Tinkle, S., Weissman, D. N., Antonini, J. M., Kashon, M. L., Salmen, R., Battelli, L. A., Willard, P. A., Hoover, M. D. & Hubbs, A. F. (2003). Efficiancy of a technique for exposing the mouse lung to particles aspirated from the pharynx. *J Toxicol Environ Health,66(15)*, 1441–1445.

Renwick, L. C., Donaldson, K. & Clouter, A. (2001). Impairment of alveolar macrophage phagocytosis by ultrafine particles. *Toxicol Appl Pharmacol,172(2),* 119–127.

Roels, H., Lauwerys, R., Buchet, J. P., Genet, P., Sarhan, M. J., Hanotiau, I., Defays, M., Bernard, A. & Stanescu, D. (1987). Epidemiologic survey among workers exposed to manganese—effects on lung, central-nervous- system, and some biological indexes. *Am J Ind Med,11(3)*, 307–327.

Schultz, J., Hohenberg, H., Plücker, F., Gärtner, E., Will, T., Pfeiffer, S., Wepf, R., Wendel, V., Gers-Barlag, H. & Wittern, K. P. (2002). Distribution of sunscreens on skin. *Adv Drug Deliv Rev,54*(Suppl 1), S1 57–S163.

Scott, R. (1997). Basic concepts of industrial hygiene. Boca Raton, LA: Lewis Publishers.

Stefaniak, A. B., Hoover, M. D., Dickerson, R. M., Peterson, E. J., Day, G. A., Breysse, P. N., Kent, M. S. & Scripsick, R. C. (2003). Surface area of respirable beryllium metal, oxide, and copper alloy aerosols and implications for assessment of exposure risk of chronic beryllium disease. *Am Ind Hyg Assoc J,64(3)*, 297–305.

Thomas, K. & Sayre, P. (2005). Research strategies for safety evaluation of nanomaterials, Part 1: evaluation the human health implications of exposure to nano- scale materials. *Toxicol Sci,87(2)*, 316–321.

Wang, H. C. & Kasper, G. (1991). Filtration efficiency of nanometer-size aerosol-particles. *J Aerosol Sci,22(1)*, 31–41.

Wang, H. C. (1996). Comparison of thermal rebound theory with penetration measurements of nanometer particles through wire screens. *Aerosol Sci Technol,24(3)*, 129–134.

Wennberg, A., Iregren, A., Struwe, G., Cizinsky, G., Hagman, M. & Johansson, L. (1991). Manganese exposure in steel smelters a health-hazard to the nervous-system. *Scand J Work Environ Health,17(4)*, 255–262.

In: Safe Nanotechnology and Progress in the Workplace... ISBN: 978-1-61761-818-5
Editor: Dominick A. Caggione

Chapter 2

PROGRESS TOWARD SAFE NANOTECHNOLOGY IN THE WORKPLACE: A REPORT FROM THE NIOSH, NANOTECHNOLOGY RESEARCH CENTER*

Department of Health and Human Services

FOREWORD

As with any new technology, the earliest and most extensive exposures to engineered nanoparticles are most likely to occur in the workplace among the workers who produce and use nanoparticles. The National Institute for Occupational Safety and Health (NIOSH) is mandated by law to conduct research and develop guidance on worker safety and health. NIOSH and its partners in other government agencies, countries, academia, industry, labor, and nongovernmental organizations have been conducting re search and developing guidance to address the occupational safety and health of workers exposed to nanomaterials.

Occupational safety and health issues of nanotechnology are complex. The types of nanomaterials and the opportunities for work place exposure continue to grow rapidly. The challenge is to effectively address the safety and health issues of nanotechnology while helping society realize the far-reaching potential benefits. NIOSH will continue to respond to this challenge.

John Howard, M.D.
Director, National Institute for Occupational Safety and Health
Centers for Disease Control and Prevention

---

*This is an edited, reformatted and augmented edition of a United States Department of Health and Human Services publication, Report 2010-104, dated November 2009.

EXECUTIVE SUMMARY

Occupational safety and health issues of nanomaterials are complex. Because engineered nanomaterials are small yet have a relatively large surface area, they may have chemical, physical, and biological properties distinctly different from larger particles of similar chemical composition. Those properties may affect the ability of the nanomaterials to reach the gas exchange regions of the lung, travel from the lung throughout the body, penetrate dermal barriers, cross cell membranes, and interact at the molecular level. The types of nanomaterials and the opportunities for work place exposure to them continue to grow rapidly. The challenge is to effectively address the safety and health issues of nanotechnology while helping society realize nanotechnology's far-reaching potential benefits.

The National Institute for Occupational Safety and Health (NIOSH) is the Federal agency responsible for conducting research and making recommendations to prevent work-related injury, illness, and death. NIOSH is mandated by the Occupational Safety and Health Act to determine whether materials in a workplace constitute any harm and to provide recommendations for preventing injury and illness.

NIOSH established the Nanotechnology Research Center (NTRC) in 2004 to coordinate and promote research in nanotechnology and to develop guidance on the safe handling of nanomaterials in the workplace. The NTRC is a virtual center in which NIOSH scientists and engineers at geographically dispersed locations are linked by shared computer networks and other technologies. This approach surmounts the logistical complications that traditionally arise when scientists and engineers collaborating on common research are not physically in the same location. In 2007 the NTRC issued *Progress Toward Safe Nanotechnology in the Workplace* (DHHS NIOSH Publication No. 2007–123). That report describes the progress of the NTRC since its inception in 2004 through 2006. This November 2009 update describes program accomplishments achieved in 2007 and 2008, and it includes summary updates from 43 intramural projects and a comprehensive extramural program. The NTRC has, with limited resources, continued to make contributions to all the steps in the continuum from hazard identification to risk management (Figure 1).

The NTRC has four goals:

- Determine whether nanoparticles and nanomaterials pose risks of work-related injuries and illnesses.
- Conduct research to prevent work-related injuries and illnesses by applying nanotechnology products.
- Promote healthy workplaces through interventions, recommendations, and capacity building.
- Enhance global workplace safety and health through national and international collaborations on nanotechnology research and guidance.

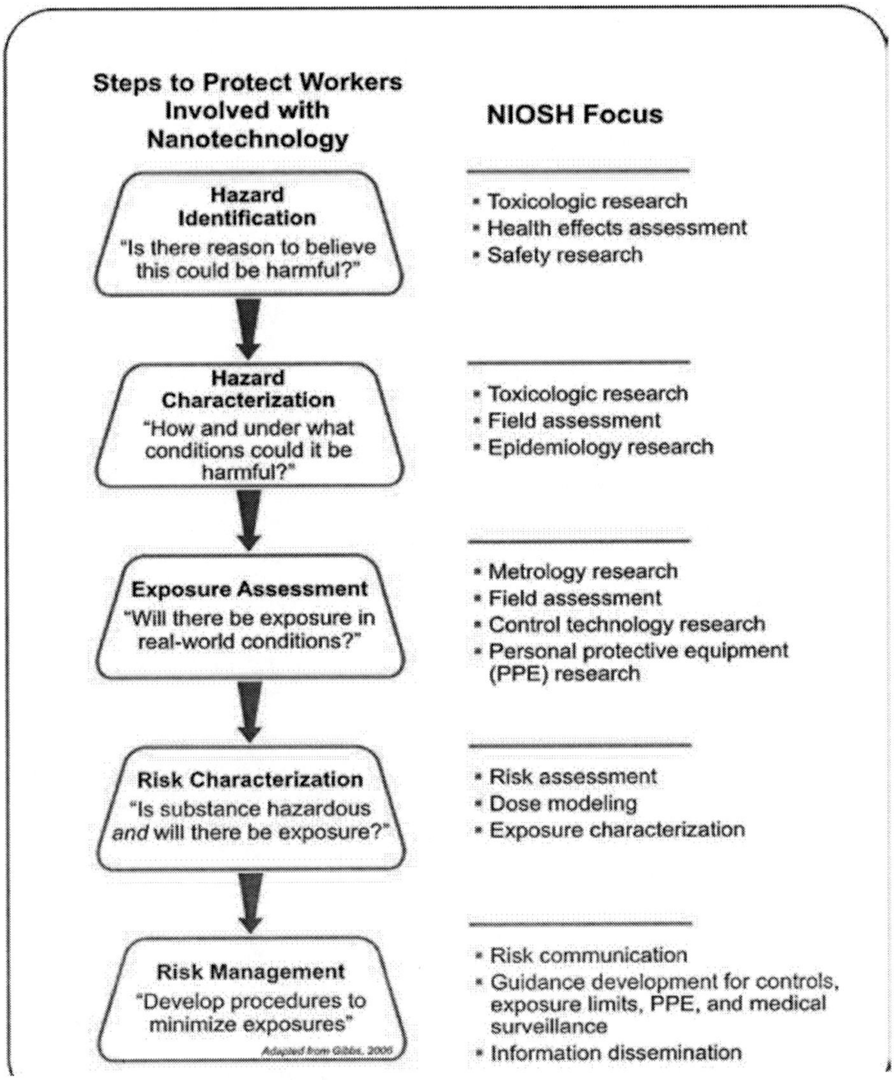

Figure 1. Steps in the continuum from hazard identification to risk management to protect workers involved with nanotechnology

As evidenced by the full report, progress continues to be made toward each of these goals. The following paragraphs present highlights of each goal.

1. Determine whether nanoparticles and nanomaterials pose risks of work-related injuries and illnesses.

Since 2004, the NTRC has pioneered research on the toxicological properties and characteristics of nanoparticles. This research is relevant for predicting whether these particles pose a risk of adverse health effects in workers. These research projects have involved characterizing occupationally relevant nanoparticles—particularly the toxicity of carbon nanoparticles. Preliminary work by the NTRC investigators demonstrated that mice exhibited harmful pulmonary effects (such as a fibrotic response) soon after low-dose aspiration exposure to specific nanotubes. The NTRC established a system for generating nanoparticle aerosol and began conducting animal inhalation studies during the summer of 2006. The inhalation studies verified the same fibrotic response seen in the aspiration studies. NTRC investigators have evaluated the potential for nanoparticles to enter the bloodstream and move to systemic tissues after being deposited in the lungs. These studies will help scientists determine whether some engineered nanomaterials pose risks to human health in the work setting. They will also help determine the mechanisms by which engineered nanomaterials operate. The results of these studies are preliminary and limited, and more research is needed to predict whether they signal a health risk to humans. The results obtained thus far support the need to address that question and provide promising leads for strategic, ongoing studies.

To gain further knowledge about exposure and control practices, the NTRC established a field team to conduct assessments of workplaces where exposure to engineered nanoparticles may occur. By partnering in 2007 and 2008 with 12 companies that produce or use engineered nanoparticles, this team collected useful information about potential worker exposures, control technologies, and risk management practices.

2. Conduct research to prevent work-related injuries and illnesses by applying nanotechnology products.

The NTRC has identified various possibilities for applying nanotechnology to occupational safety and health, including the application of this technology in fabricating more efficient filters, sensors, and protective clothing. The NTRC and the Office of Extramural Programs at NIOSH are coordinating projects with academia and the private sector.

3. Promote healthy workplaces through interventions, recommendations, and capacity building.

Nationally and internationally, the NTRC has delivered a wide range of presentations on occupational safety and health issues associated with nano- technology. These have included presentations to scientific conferences, trade associations, and professional associations; meetings of government agencies and nongovernmental organizations (NGOs); and special panels convened by government agencies, NGOs, and professional associations. Since its inception, the NTRC has published more than 170 papers in peer-reviewed scientific literature and provided a broad range of information and guidance. In addition, the NTRC has provided seminal guidance for workers and employers in nanotechnology through the following documents:

- *Interim Guidance for Medical Screening and Hazard Surveillance for Workers Potentially Exposed to Engineered Nanoparticles,* DHHS NIOSH Publication No. 2009–116

- *Approaches to Safe Nanotechnology: Managing the Health and Safety Concerns Associated with Engineered Nanomaterials,* DHHS NIOSH Publication No. 2009–125
- *Safe Nanotechnology in the Workplace: An Introduction for Employers, Managers, and Safety and Health Professionals,* DHHS NIOSH Publication No. 2008–112
- The Nanotechnology Field Research Team Update, DHHS NIOSH Publication No. 2008–120
- *NIOSH Nanotechnology Field Research Effort Fact Sheet,* DHHS NIOSH Publication No. 2008–12 1
- *NIOSH Nanotechnology Metal Oxide Particle Exposure Assessment Study*, DHHS NIOSH Publication No. 2008–122
- Draft *Strategic Plan for NIOSH Nanotechnology Research: Filling the Knowledge Gaps*, published February 2008 [www.cdc.gov/niosh/topics/ nanotech/strat_plan A.html]

4. Enhance global workplace safety and health through national and international collaborations on nanotechnology research and guidance.

The NTRC has established several national and international collaborations to advance understanding of occupational safety and health for nanotechnology workers. The NTRC participates in the National Nanotechnology Initiative (NNI) and has contributed to the nanotechnology strategic plan for the Nation through the Nanotechnology Environmental Health Implications (NEHI) working group. Occupational safety and health has been a chief priority of the NEHI effort, and the NIOSH strategic research plan and activities are addressing most of the major issues in the NEHI plan.

The NTRC is part of the U.S. leadership on the International Organization for Standardization (ISO) TC 229, Nanotechnology Working Group on Health, Safety, and the Environment. NIOSH led the development and publication of ISO TR/ 12885:2008, *Health and Safety Practices in Occupational Settings Relevant to Nanotechnologies*. The NTRC also works with the World Health Organization collaborating centers on global projects of information dissemination and communication. NIOSH co-led the World Health Organization project on best practices for safe handling of nanomaterials.

The NTRC has collaborated with the Organization for Economic Cooperation and Development (OECD) to build cooperation, coordination, and communication among the United States and 30 OECD member countries (including the European Union) and with more than 180 nonmember economies. NIOSH chaired the OECD Working Party for Manufactured Nanomaterials Steering Group 8, exposure measurement and exposure mitigation.

NIOSH cosponsored the 3rd International Symposium on Nanotechnology, Occupational and Environmental Health held in Taiwan in 2007 and serves on the planning committee for the 4th international conference scheduled for August 2009 in Helsinki, Finland. In addition, the NTRC is collaborating with the U.S. Environmental Protection Agency, the Occupational Safety and Health Administration, and other government agencies to obtain and evaluate exposure and good work practice information.

Critical Topic Areas

The NTRC research program has identified 10 critical topic areas for understanding the potential health risks and developing and disseminating recommendations:

1. Toxicity and internal dose
2. Measurement methods
3. Exposure assessment
4. Epidemiology and surveillance
5. Risk assessment
6. Engineering controls and PPE
7. Fire and explosion safety
8. Recommendations and guidance
9. Communication and information

10. Applications

Table 1. Alignment of the Four Strategic Goals and Ten Critical Research Areas of the NIOSH NTRC with the Priority Environmental Health and Safety Research Needs of the NNI*

NIOSH NTRC strategic goals	Determine if nanoparticles and nanomaterials pose risks for work-related injuries and illnesses	Conduct research on applying nanotechnology to the prevention of work-related injuries and illnesses[1]	Promote healthy workplaces through interventions, recommendations, and capacity building	Enhance global workplace safety and health[1]
Toxicity and internal dose	✓[3], A2[4], B1, B2, B3, B4, B5			✓
Measurement methods	✓, A1, A2, A3, A4, A5, B2	✓	✓, E1	✓
Exposure assessment[2]	✓, A1, A2, D1, D4, D5, E1, E2		✓, E2	✓
Epidemiology and surveillance	✓, D1, D2, D3, D4, E4		✓, E4	✓
Risk assessment	✓, A2, B1, B2, D4, E3		✓, E3, E5	✓
Engineering controls and PPE	✓, D5	✓	✓, E1	✓
Fire and explosion safety	✓, A2			✓
Recommendations and guidance			✓, E1, E5	✓
Communication and information			✓, E1, E5	✓
Applications		✓		✓

[1] The NNI EHS plan does not address applications or global activities pertaining to environmental health and safety of nanomaterials. However the NIOSH efforts described herein are consistent with the overall NNI strategy which does identify international opportunities for collaboration.

[2] NIOSH is the lead agency for the NNI EHS Strategic area on human and environmental exposure assessment.

[3] A ✓ indicates that a goal is addressed by projects within the NIOSH critical research area

[4] Alpha-numeric identifications indicate alignment of the NIOSH critical research area with the specific NNI priority research needs (see Appendix A).

By working in these 10 critical areas, NIOSH has comprehensively begun addressing the information and knowledge gaps necessary to protect workers and responsibly move nanotechnology forward so that its far-reaching benefits may be realized.

Coordination with the National Nanotechnology Initiative

NIOSH participates in the NNI Strategy for Nanotechnology Environmental Health and Safety Research. Table 1 shows the alignment of the four NIOSH NTRC strategic goals and the 10 NTRC critical research areas with the NNI priority environmental health and safety research needs.

The research being conducted by the NIOSH NTRC continues to be funded by redirecting existing NIOSH programmatic funds: $4.4 million in FY 2007 and $ 5.7 million in FY 2008. A more comprehensive research program specific to nanomaterials has been difficult to implement because of this budgetary constraint. However, even with the budgetary constraints, NIOSH investigators have continued to build an evidence-based strategy for providing safe nano- technology in the workplace.

ABBREVIATIONS

AIHA	American Industrial Hygiene Association
ACGIH	American Conference of Governmental Industrial Hygienist
AM	alveolar macrophage
AP- 1	activator protein 1
APR	air–purifying respirators
ASME	American Society of Mechanical Engineers
ASSE	American Society of Safety Engineers
ASTM	American Society for Testing and Materials
ATP	adenosine triphosphate
BAL	bronchoalveolar lavage
CARB	California Air Resources Board
CDC	Centers for Disease Control and Prevention
CNF	carbon nanofiber
CNT	carbon nanotube
COPD	Chronic obstructive pulmonary disease
CPSC	Consumer Product Safety Commission
DEP	diesel exhaust particulate
DEER	Diesel Engine– Efficiency and Emissions Research
DEPE	diesel exhaust particulate extract
DERT	Developmental Engineering Research Team
DHHS	U.S. Department of Health and Human Services
DIC	different interference confocal
DNA	dioxyribonucleic acid
DOE	U.S. Department of Energy
DPF	diesel particulate filter
EC–SOD	extracellular superoxide dismutase
EHS	Environmental Health and Safety
EPA	U.S. Environmental Protection Agency
ESD	electron spin resonance

ESLI	end-of-service life indicator
ESR	erythrocyte sedimentation rate
FASEB	Federation of American Societies for Experimental Biology
FCA–SS	flux core arc–stainless steel
FDA	U.S. Food and Drug Administration
FISH	fluorescent in situ hybridization
GMA–MS	gas metal arc–mild steel
HEI	Health Effects Institute
HEPA	high-efficiency particulate air
HHE	Health Hazard Evaluation
HPS	Health Physics Society
IANH	International Alliance for NanoEHS Harmonization
IDLH	immediately dangerous to life and health
IEC	International Electrotechnical Commission
IOM	Institute of Occupational Medicine
ILO	International Labor Organization
ILSI	International Life Sciences Institute
ISO	International Organization for Standardization
IRSST	Institut de recherché Robert–Sauvè en santé et en sécurité du travail
kW	kilowatt
LDH	lactate dehydrogenase
LLC	limited liability company
LM	Listeria monocytogenis
LNW	long nanowire
MAPKs	mitogen–activated protein kinase
MINChar	Minimum Information Needed for Characterization of Nanomaterials
MMA–HS	manual metal arc–hard surfacing
MOUDI	micro–orifice uniform deposit impactor
MPPD	multipath particle deposition
MWCNT	multi-walled carbon nanotube
NADPH	nicotinamide adenine dinucleotide phosphate
NASA	National Aeronautics and Space Administration
NATO	North Atlantic Treaty Organization
NCER	National Center for Environmental Research
NEAT	nanoparticle emission assessment technique
NEHI	Nanotechnology Environmental and Health Implications
NFkB	nuclear factor kB
NGO	non-governmental organization
NIEHS	National Institute of Environmental Health Sciences
NIH	National Institutes of Health
NIL	Nanoparticle Information Library
NIOSH	National Institute for Occupational Safety and Health
NIST	National Institute of Standards and Technology
NO	nitric oxide
nm	nanometer(s)
NNI	National Nanotechnology Initiative

NORA	National Occupational Research Agenda
NS	nanosphere
NSC	National Safety Council
NSET	Nanoscale Science, Engineering, and Technology
NSF	National Science Foundation
NSTC	National Science and Technology Council
NTRC	Nanotechnology Research Center
NSC	National Safety Council
OEP	Office of Extramural Programs
OECD	Organization for Economic Cooperation and Development
OEP	Office of Extramural Programs
OPN	osteoponin
OSHA	U.S. Occupational Safety and Health Administration
PEL	permissible exposure limit
PMN	polymorphonuclear neutrophils
POSS	polyhedral oligomeric silsesquioxanes
PPE	personal protective equipment
r2p	Research to Practice
RFA	request for application
RNS	reactive nitrogen species
ROS	reactive oxygen species
SBIR	small business innovation research
SMPS	scanning mobility particle sizer
SNW	short nanowire
SRM	standard reference material
STEL	short term exposure limit
SWCNT	single-walled carbon nanotube
TC	technical committee
TEM	transmission electron microscopy
TGF–ß	transforming growth factor beta
TiO_2	titanium dioxide
TR	technical report
µg/m3	microgram per cubic meter
UK	United Kingdom
USAF	U.S. Air Force
VEGF	vascular endothelial growth factor
Wc–Co	tungsten carbide
WHO	World Health Organization
WT	wild type

INTRODUCTION

Background

The NIOSH Nanotechnology Research Center (NTRC) was established in 2004 to coordinate and promote research in nanotechnology and develop guidance on the safe handling of nanomaterials in the workplace. Since its inception, the NTRC has published more than 170 papers in peer-reviewed scientific literature and provided a broad range of information and guidance (see 2007 Progress Report http://www.cdc.gov/niosh/docs/2007-123/ and Table 2). Publications in the areas of hazard identification and characterization, exposure assessment, risk assessment, and risk management have provided a framework for addressing hazards and risks from engineered nanoparticles. Although NTRC has contributed significantly to the understanding of health risks to nanoparticles, further research and subsequent development and dissemination of interim guidance efforts are limited by a lack of funding. NIOSH has to the best of its ability allocated limited resources to competing priorities and redirected funds internally over the past 4 years to build modest increases in support of the NIOSH Nanotechnology Research Program. New infusions of funding beyond the current NIOSH budget, specifically directed toward the program, would be needed for significant expansion of research to address the following research gaps:

- Although NIOSH toxicology studies have provided better understanding of the ways in which some types of nanoparticles may enter the body and interact with the body's organ systems, the breadth and depth of such research efforts have been limited to a few nanoparticle types. More types of nanoparticles need to be assessed for characteristics and properties relevant for predicting potential health risks.
- NIOSH field investigators have assessed exposure to engineered nanoparticles in some workplaces, but little data exist on the extent and magnitude of exposure to other types of nanoparticles in workplaces that manufacture or use nanomaterials, nanostructures, and nanodevices.
- NIOSH guidance is a first step toward controlling nanoparticles in the workplace; however, more research is needed on the efficacy and specificity of engineering and work practice control measures. NIOSH needs support to conduct more field investigations.
- The number of workers involved with the research, development, production, and use of engineered nanomaterials is increasing, but there is a lack of specific guidance for occupational health surveillance. NIOSH needs support to conduct research on the short- and long-term health risks in nanotechnology workers. This may involve conducting large-scale prospective epidemiologic studies and establishing exposure registries.
- The utility of nanotechnology to support development of new technologies (such as sensors, more efficient filters, and better protective materials) that can enhance the protection of workers requires further research and development and can be advanced by additional resources.

The NIOSH NTRC has advanced the scientific knowledge in understanding the possible health risks of engineered nanoparticles as demonstrated by the ongoing projects beginning on page 15. Table 2 (see Appendix B) lists the projects by number and critical topic areas researched (e.g., toxicity and internal dose, measurement methods, exposure assessment). These projects are also listed in Appendix A for each critical topic area.

Partnerships and Collaborations

Integral to completing this research are the partnerships and collaborations NIOSH has forged with countries, academia, industry, labor, other government agencies, and nongovernmental organizations. These collaborations have provided expertise and resources critical for successfully completing research and for developing and disseminating health and safety information on engineered nanoparticles. The following are some of the ongoing NIOSH NTRC partnerships and collaborations in nanotechnology research:

Universities

- The Ohio State University, evaluating workplaces, controls, and practices in the nanocomposites industry
- University of Cincinnati, translating nanotechnology research into occupational and environmental health and safety activities
- University of Dayton Research Institute, understanding and improving the control of carbon nanotubes
- University of Iowa, measuring airborne concentrations of ultrafine particles
- University of Nevada at Reno (with Altair Nano), understanding and improving the control of ultrafine metal oxides and engineered nano- materials
- University of Massachusetts at Lowell, collaborating on the development of good practices for handling nanomaterials in research laboratories and developing exposure assessment methods
- University of Minnesota, measuring airborne concentrations of ultra- fine particles
- University of Montana, evaluating the toxicity of nanowires
- University of Montreal, assessing and characterizing nanoparticles in the workplace
- University of Pittsburgh, evaluating the toxicity of nanomaterials
- University of Rochester, measuring the ability of nanoparticles to generate radical species
- University of Wisconsin, Madison, developing exposure assessment strategies and characterizing nanoparticles in friction stir welding and micromachining
- Virginia Tech University (with Luna Nano Works), understanding and improving the control of fullerines and other engineered nanoparticles
- West Virginia University Nanotechnology Center, evaluating the toxicity of metal oxide nanowires
- West Virginia University Medical School, evaluating the effects of pulmonary exposure to nanoparticles, including the influence of exposure to titanium dioxide (TiO_2) on systemic microvascular function

Safety and Health Organizations and Expert Groups

- American Industrial Hygiene Association (AIHA), participating on working group activities and information exchange and dissemination for nanoparticle characterization and control
- American Conference of Governmental Industrial Hygienists (ACGIH)
- American Society of Safety Engineers (ASSE)
- Hamner Institutes for Health Sciences, modifying software for use in lung dosimetry modeling
- Health Physics Society (HPS), participating in working group activities and information exchange and dissemination for nanoparticle characterization and control
- International Safety Equipment Association, collaborating on the respirator filter media work being conducted at the University of Minnesota to assess the filtration efficiency against nanoparticles
- International Alliance for Nanotoxicology Harmonization (IANH), establishing protocols for reproducible toxicological testing of nanomaterials in both cell cultures and animals
- National Safety Council (NSC)
- The initiative on Minimum Information Needed for Characterization of Nanomaterials (MINChar), on encouraging adoption of a minimum set of physical and chemical material characterization parameters in nanotoxicology studies
- Institute of Occupational Medicine (IOM), Edinburgh, Scotland, revising rat lung dosimetry models to account for particle size-specific clearance and retention.

Federal and Other Government Agencies

- Environmental Protection Agency (EPA)
- Occupational Safety and Health Administration (OSHA), evaluating the effectiveness of workplace control measures
- Consumer Product Safety Commission (CPSC)
- National Aeronautics and Space Administration (NASA), evaluating the toxicity of single-walled carbon nanotubes (SWCNTs)
- U.S. Air Force
- U.S. Army
- Department of Energy (DOE)
- National Institute of Environmental Health Sciences/ National Institute of Health and the Department of Defense on nanotoxicology
- National Institute of Standards and Technology (NIST), evaluating the toxicity of nanoparticles and developing nanoscale reference materials
- National Nanotechnology Initiative (NNI) through active participation in NNI's strategic and budget planning processes, and coordination of research into occupational safety and health of nanotechnology with other government agencies
- National Toxicology Program, participating in the conduct of field assessments and toxicological research

- Oak Ridge National Laboratory, evaluating the pulmonary toxicity of nanoparticles
- Institut de Recherche Robert-Sauve en Santé et en Securite du Travail (IRSST), exchanging information on approaches to safe nanotechnology
- Karolinska Institute, evaluating the effect of nanomaterials on immune cells
- U.S. Nuclear Regulatory Commission, exchanging information on emerging application and health and safety issues for nanotechnology in the nuclear arena
- World Health Organization (WHO)
- Organization for Economic Cooperation and Development (OECD),collaborating on exposure measurement and exposure mitigation
- International Labour Organization (ILO), developing information on nanotechnology for a revised edition of the Encylopaedia of Occupaitonal Health and Safety
- International Organization for Standardization (ISO) and the International Electrotechnical Commission (IEC), developing international standards for terminology, characterization, and safety practices for nanotechnology

Nanotechnology Industries

- DuPont and the Nanoparticle Occupational Safety and Health industry consortium on issues related to the measurement of nanoparticles and the efficiency of filtration materials for engineered nanoparticles
- Mitsui Co. Inc., evaluating the bioactivity of multi-walled carbon nano- tubes (MWCNTs).
- IBM, evaluating the bioactivity of silicon nanowires
- Luna Nano Works, understanding and improving the control of fullerines and other engineered nanoparticles
- Altair Nano, understanding and improving the control of ultrafine metal oxides and engineered nanomaterials
- Quantum Sphere, understanding and improving the control of nanoscale metals and metal oxides
- QD Vision, understanding and improving the control of quantum dots
- NanoMech LLC (Fayetteville, Arkansas), collaborating on a proposed EPA phase I SBIR project related to containment of airborne nanoparticles

Collaborative Workshops

- In FY 2005, NIOSH cosponsored the first Symposium on OccupationalHealth Implications of Nanomaterials in Buxton, United Kingdom
- In FY 2006, NIOSH cosponsored the second Symposium on Occupational Health Implications of Nanomaterials in Minneapolis, MN.
- In 2007,
    - NIOSH collaborated with the International Aerosol Research As-sembly and the American Association for Aerosol Research to hold the Second International Symposium on Nanotechnology and Occupational Health in conjunction with the international aerosol conference in St. Paul, MN.

- NIOSH and the University of Cincinnati cosponsored the International Conference on Nanotechnology Occupational and Environmental Health and Safety: Research to Practice, in Cincinnati, OH.
- NIOSH convened a collaborative workshop that included representatives from government, academia, labor, and industry in Washington, DC, to review a draft document developed by NIOSH and a cross-agency work group, titled *Interim Guidelines on Medical Screening of Workers Potentially Exposed to Engineered Nanoparticles.*
- NIOSH cosponsored the Third Symposium on Occupational Health Implications of Nanomaterials in Taipei, Taiwan.

- In 2008,
  - NIOSH served as the agency lead to organize the program for the National Nanotechnology Initiative Workshop on Human and Environmental Exposure Assessment held February 2009, in Bethesda, MD.
  - NIOSH co-sponsored a National Institute of Standards and Technology Workshop on Enabling Standards for Nanomaterial Characterization in Gaithersburg, MD.
  - NIOSH co-sponsored The Second International Conference on Nanotechnology in Zurich, Switzerland.
  - NIOSH co-sponsored the Organization for Economic Co-operation and Development workshop on Exposure Assessment and Exposure Mitigation in Frankfurt, Germany.
  - NIOSH participated in the planning of the Fourth International Conference on Nanotechnology—Occupational and Environmental Health held in August 2009, in Helsinki, Finland.

NIOSH Extramural Nanotechnology Research Activities

The NIOSH Office of Extramural Programs (OEP) manages the competitive process for awarding occupational safety and health grants and cooperative agreements to the research community outside the Institute. This process includes peer review, program relevance, and priorities from the National Occupational Research Agenda (NORA), the NIOSH r2p initiative, congressional mandates, and sector, or cross-sector areas of the NIOSH Program Portfolio (http://www.cdc.gov/niosh/programs).

From 2001 to 2009, the Office of Extramural Programs (OEP) has funded nanotechnology research through Occupational Safety and Health Research Program Announcements (R01) and Small Business Innovation Research Grants (R43/44). Since FY 2005, OEP has also participated in two joint Requests for Applications (RFAs) for Nanotechnology Research Grants Investigating Environmental and Human Health Issues. The US Environmental Protection Agency's National Center for Environmental Research (NCER) and the National Science Foundation (NSF) participated in FY 2005. The National Institute of Environmental Health Sciences (NIEHS) joined in FY 2006. Funding was available to support Research (R01) grants for three years and Exploratory (R21) grants for two year.

In FY 2007, NIOSH/OEP participated in RFA-ES-06-008 Manufactured Nano- materials: Physico-chemical Principles of Biocompatibility and Toxicity. This RFA was jointly sponsored by NIEHS, EPA, and NIOSH.

In FY 2009, nanotechnology related research proposals submitted to standing program announcements are being considered for funding by NIOSH/OEP.

Table 2. NTRC Projects in the 10 Critical Topic Areas for 2007–2008

Project number	Toxicology and Internal dose	Measurement methods	Exposure assessment	Epidemiology and surveillance	Risk assessment	Engineering controls and PPE*	Fire and explosion	Recommen-dations/ guidance	Communi-cation and information	Applications
1	X	X	X	X	X	X	X	X	X	X
2			X							
3	X									
4	X									
5	X									
6	X									
7	X									
8	X									
9	X									
10	X									
11	X									
12	X									
13	X									
14	X									
15	X									
16	X									
17	X									
18	X									
19	X									
20	X									
21	X									
22		X								
23		X	X				X			
24		X	X							
25		X	X							
26		X								
27		X								
28		X								
29		X								
30		X								
31		X	X							
32		X								
33		X	X							
34			X							
35				X						X
36					X					
37						X				
38						X				
39						X				
40									X	
41			X						X	
42										X
43			X		X			X	X	

*PPE=personal protective equipment

To date, NIOSH/OEP has committed about $5.3 million dollars to research on applications and implications of nanotechnology. Summaries of the projects funded by NIOSH/OEP are included in Appendix B. NIOSH/OEP plans to continue collaborative efforts with EPA/NCER, NSF, NIH/NIEHS, and other international agencies to support nanotechnology research with occupational safety and health implications. OEP will continue to confer with the NIOSH Nanotechnology Research Center regarding issues, gaps, and future directions.

Project-Specific Progress Reports

Project 1. Nanotechnology Research Center Coordination

Principal Investigator: Charles Geraci, Ph.D.
Project Duration: FY 2006–2010

Critical Topic Areas: (1) toxicity and internal dose, (2) measurement methods, (3) exposure assessment, (4) epidemiology and surveillance, (5) risk assessment, (6) engineering controls and personal protective equipment, (7) fire and explosion safety, (8) recommendations and guidance, (9) communication and information, and (10) applications

Accomplishments and Research Findings

- Coordinated the NIOSH Nanotechnology Research Program across 10 critical areas identified in the strategic research plan developed by the NTRC in 2005 to identify and prioritize nanotechnology research.
- Sponsored and provided strategic direction to the Nanotechnology Field Research Team, which conducted 14 site visits in the reporting period.
- Fostered collaboration among investigators across the entire NIOSH Nanotechnology Research Program.
- Developed and maintained successful interagency agreements with
    - Occupational Safety and Health Administration
    - Consumer Product Safety Commission
    - National Toxicology Program
- Fostered collaboration in nanotechnology research with
    - AIHA
    - ASSE
    - NSC
    - U.S. Air Force
    - U.S. Army
    - NASA
    - DOE
    - International Alliance for NanoEHS Harmonization

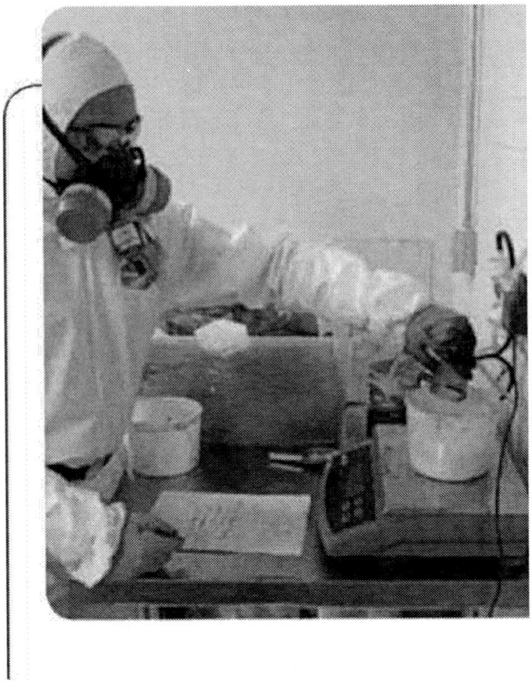

Benchtop weighing of CNT. *Image courtesy of Mark Methner*, NIOSH

    - University of Massachusetts, Lowell
    - The Ohio State University
    - University of Dayton Research Institute
- Represented NIOSH on several national and international working groups on nanotechnology, safety, and health.
- Presented overviews of the NIOSH Nanotechnology Research Program to numerous organizations.
- Provided briefings to the President's Council of Advisors on Science and Technology.

Publications and Abstracts

- Hoover MD, Stefaniak AB, Day GA, Geraci CL [2007]. Exposure assessment considerations for nanoparticles in the workplace. In: Monteiro-Riviere NA, Tran CL, eds. Nanotoxicology: characterization, dosing, and health effects. New York: Informa Healthcare USA Inc., pp. 71–83.
- Kuempel ED, Geraci CL, Schulte PA [2007]. Risk assessment approaches and research needs for nanomaterials: an examination of data and information from current studies. In: Simeonova PP, Opopol N, Luster MI, eds. Nanotechnology: toxicological issues and environmental safety, NATO security through science series book. New York: SpringerVerlag, pp. 119–145.
- NIOSH [2008]. Safe nanotechnology in the workplace. An introduction for employers, managers, and safety and health professionals. Cincinnati, Ohio: U.S.

Department of Health and Human Services, Centers for Disease Control and Prevention, National Institute for Occupational Safety and Health, DHHS (NIOSH) Pub No. 2008–112.

- NIOSH [2008]. The nanotechnology field research team update. Cincinnati, Ohio: U.S. Department of Health and Human Services, Centers for Disease Control and Prevention, National Institute for Occupational Safety and Health, DHHS (NIOSH) Pub No. 2008–120.
- NIOSH [2008]. NIOSH nanotechnology field research effort fact sheet. Cincinnati, Ohio: U.S. Department of Health and Human Services, Centers for Disease Control and Prevention, National Institute for Occupational Safety and Health, DHHS (NIOSH) Pub No. 2008–121.
- NIOSH [2008]. NIOSH nanotechnology metal oxide particle exposure assessment study. Cincinnati, Ohio: U.S. Department of Health and Human Services, Centers for Disease Control and Prevention, National Institute for Occupational Safety and Health, DHHS (NIOSH) Pub. No. 2008–122.
- NIOSH [2008]. Draft Strategic plan for NIOSH nanotechnology research: filling the knowledge gaps. Cincinnati, Ohio: U.S. Department of Health and Human Services, Centers for Disease Control and Prevention, National Institute for Occupational Safety and Health, February 2008 [www.cdc.gov/niosh/topics/nanotech /strat_planA.html].
- Schulte PA, Geraci CL, Zumwalde RD, Hoover MD, Kuempel ED [2008]. Occupational risk management of nanoparticles. J Occup Environ Hyg 5(4):239–249.
- Schulte PA, Trout D, Zumwalde R, Kuempel E, Geraci CL, Castranova V, Mundt D, Mundt K, Halperin W [2008]. Options for occupational health surveillance of workers potentially exposed to engineered nanoparticles: state of the science. J Occup Med *50*(5):517–526.
- Schulte PA, Geraci CL, Zumwalde R, Hoover MD, Castranova VA, Kuempel E, Murashov V, Vainio H, Savolainen K [2008]. Sharpening the focus on occupational safety and health in nanotechnology. Scand J Work Environ Health *34*(6):471–478.
- NIOSH [2009]. Interim guidance for medical screening and hazard surveillance for workers potentially exposed to engineered nanoparticles. Cincinnati, Ohio: U.S. Department of Health and Human Services, Centers for Disease Control and Prevention, National Institute for Occupational Safety and Health, DHHS (NIOSH) Pub. No. 2009–116.
- NIOSH [2009]. Approaches to safe nanotechnology: managing the health and safety concerns with engineered nanomaterials. Cincinnati, Ohio: U.S. Department of Health and Human Services, Centers for Disease Control and Prevention, National Institute for Occupational Safety and Health, DHHS (NIOSH) Pub. No. 2009–125.

Invited Presentations

- Geraci CL [2007]. Nanotechnology: an overview of health and safety issues "implications and applications." Ohio Safety Congress and Expo, Columbus, OH, March 20.

- Geraci CL [2007]. The NIOSH Nanotechnology Research Program: occupational implications and applications. General Electric nanotechnology summit, Miskayuna, NY, April 18.
- Geraci CL [2007]. Nanotechnology: using basic risk management concepts to address difficult questions. Society for Chemical Hazard Communication spring meeting, San Antonio, TX, April 25.
- Geraci CL [2007]. The NIOSH Nanotechnology Research Program: occupational implications and applications. Northeastern University panel on nanotechnology and policy, Boston, MA, May 1.
- Geraci CL [2007]. Nanomaterials in the workplace: applying basic risk management principles. American Society of Safety Engineers webinar, May 15.
- Geraci CL [2007]. Nanotechnology: an international perspective from NIOSH. Education Safety Association of Ontario annual conference, Ontario, Canada, May 31.
- Geraci CL, Hoover MD [2007]. Exposure assessment and sampling issues for nanotoxicology and safe handling of nanoparticles in the workplace. American Industrial Hygiene Association conference and exposition, roundtable 201: Toxicology and exposure assessment issues for nanotechnology: an update for industrial hygienists, Philadelphia, PA, June 2–7.
- Geraci CL [2007]. One year later: an update of the NIOSH Nanotechnology Research Program. American Industrial Hygiene Association conference and exposition, roundtable 218: Ask the experts: an update on the NIOSH Nanotechnology Research Program, Philadelphia, PA, June 2–7.
- Geraci CL [2007]. Building OS&H capability in nanomaterial manufacturing: an example of "research to practice." American Society of Safety Engineers conference and exposition, Orlando, FL, June 25.
- Geraci CL [2007]. Meeting the challenge: an update of the NIOSH Nanotechnology Research Program. ORC Worldwide Occupational Safety and Health joint meeting, Washington, DC, August 8.
- Geraci CL, Methner MM, Dunn KH, Coe-Sullivan S [2007]. Occupational safety and health field study of workers potentially exposed to quantum dot nanomaterials. International symposium on nano- technology: occupational and environmental health, Taipei, Taiwan, August 29.
- Geraci CL [2007]. Occupational risk management for nanomaterials along the product life cycle. International symposium on nanotechnology: occupational and environmental health, Taipei, Taiwan, September 1.
- Geraci CL [2007]. International collaboration opportunities in nano- technology research. International symposium on nanotechnology: occupational and environmental health, Taipei, Taiwan, September 1.
- Geraci CL [2007]. Progress in addressing the occupational safety and health concerns of nanotechnology. President's Council of Advisors on Science and Technology, briefing panel, Washington, DC, September 11.
- Geraci CL [2007]. Workplace risk management of nanomaterials: challenges and lessons learned. NanoTX'07, Dallas, TX, October 3.

- Geraci CL [2007]. Nanotechnology: the new workplace, industrial hygiene implications. Professional conference on industrial hygiene, Louisville, KY, October 17.
- Geraci CL [2007]. Nanotechnology: the new workplace, Nano 101 for the industrial hygienist. Professional conference on industrial hygiene, Louisville, KY, October 23.
- Geraci CL [2007]. The NIOSH Nanotechnology Research Program. Professional conference for industrial hygiene, Louisville, KY, October 23.
- Geraci CL [2007]. Nanotechnology: the new workplace, workplace exposure assessment challenges. American College of Occupational Medicine State-of-the-Art Conference/International Conference on Health Care Worker Health, Vancouver, British Columbia, Canada, October 27.
- Geraci CL [2007]. Safe development of nanotechnology. Massachusetts governor's panel on nanotechnology-office of technical assistance and technology, Boston, MA, November 15.
- Geraci CL, Hoover MD [2007]. A risk assessment and control approach, teleweb on nanotechnology health and safety: case studies in the occupational setting. American Conference of Governmental Industrial Hygienists, December 4.
- Hoover MD, Geraci CL [2008]. Nanotoxicology issues for assessing and managing occupational exposure risks for nanoparticles in the workplace. American Industrial Hygiene Association distance learning program, January 31.
- Geraci CL [2008]. Occupational safety and health issues and nanotechnology. Education Resource Committee program directors meeting, Galveston, TX, February 10.
- Schulte PA [2008]. Occupational safety and health issues and nanotechnology. Food and Drug Law meeting, Washington, DC, February 18 and 19.
- Geraci CL [2008]. Progress in addressing the occupational safety and health concerns of nanotechnology. Meeting with Lockheed Martin, Baltimore, MD, February 1.
- Geraci CL [2008]. Progress in addressing the occupational safety and health concerns of nanotechnology. Meeting with the New Jersey Technology Council, Rutgers University, Camden, NJ, March 4.
- Geraci CL [2008]. Nanotechnology and occupational safety and health: What do we know? SESHA annual high technology environmental health and safety symposium, Portland, OR, March 27.
- Geraci CL, Methner MM, Pearce TA, Hodson LL [2008]. Nanotechnology for industrial hygienists. American Industrial Hygiene Association—Central Ohio section meeting, Columbus, OH, April 22.
- Geraci CL, Hoover MD, Heidel DS, Pearce TA, Methner MM, Hodson LL [2008]. Nanoparticles in the workplace: assessing and managing risks. American Industrial Hygiene Association conference and exposition, Minneapolis, MN, May 31–June 5.
- Geraci CL [2008]. Update on NIOSH research on the measurement of nanoparticles in the workplace. American Industrial Hygiene Association conference and exposition, roundtable 218: Ask the experts: an update on NIOSH Nanotechnology Research Program, Minneapolis, MN, May 31–June 5.
- Geraci CL [2008]. NIOSH guidance on nanotechnology: protecting our workers environmental, health, and safety issues in nanomaterials workshop. Meeting

sponsored by The American Ceramic Society/National Institute of Standards and Technology, Gaithersburg, MD, June 8–10.

- Geraci CL [2008]. Nanotechnology and occupational safety and health: What do we know? International conference on nanotechnology for the forest products industry, St. Louis, MO, June 25.
- Geraci CL [2008]. The NIOSH nanotechnology field research team—a brief summary and update of activities. DOE nanoscale science research centers symposium: safe handling of engineered nanoscale materials, teleseminar to Argonne, IL, July 9.
- Geraci CL [2008]. Engineering controls: the continuum of prevention through design, control banding, and engineering controls. OECD Working Party on Manufactured Nanomaterials Workshop on Exposure Assessment and Exposure Mitigation, Frankfurt, Germany, October 20.
- Geraci CL [2008]. Nanotechnology: real world challenges for the industrial hygienist. Professional Conference on Industrial Hygiene. Tampa, FL, November 11.
- Geraci CL [2008].]. Nanotechnology: real world challenges for the industrial hygienist. California Industrial Hygiene Council, San Diego, CA, December 8.
- Methner MM, Hodson LL and Geraci CL [2008] Nanoparticle emission assessment technique (NEAT) used by NIOSH for Identifying Sources and Releases of Engineered Nanoparticles. Ohio Valley American Industrial Hygiene Association section, West Chester, OH, December 16.

Project 2. Nanotechnology Safety and Health Research Coordination

Principal Investigator: Vincent Castranova, Ph.D.
Project Duration: FY 2004–2008

Critical Topic Area: toxicology and internal dose

Accomplishments and Research Findings

- Coordinated the NIOSH nanotoxicology program.
- Fostered collaboration among investigators in the nanotoxicology program.
- Fostered collaboration in nanotechnology research with
  - The Consumer Products Safety Commission
  - International Alliance for NanoEHS Harmonization
  - Mitsui & Co.
  - West Virginia University
  - NASA
  - University of Rochester
  - University of Montana
  - University of Pittsburgh
  - Karolinska Institute

- Represented NIOSH during involvement with several national and international working groups on nanotechnology safety and health.
- Represented NIOSH during involvement with the National Nanotechnology Initiative nanotechnology environmental and health implication strategic planning group.
- Presented overviews of the NIOSH nanotoxicology program to numerous organizations.

Publications and Abstracts

- Mossman BT, Borm PJ, Castranova V, Costa DL, Donaldson K, Kleeberger SR [2007]. Mechanisms of action of inhaled fibers, particles, and nanoparticles in lung and cardiovascular diseases. Part Fibre Toxicol *4*:4.
- Leonard SS, Castranova V, Chen BT, Schwegler-Berry D, Hoover M, Piacitelli C, Gaughan D [2007]. Particle size dependent generation from wildland fire smoke. Toxicol *236*:103–113.
- Balbus JM, Maynard AD, Colvin VL, Castranova V, Daston GP, Denison RA, Dreher KL, Goering PL, Goldberg AM, Kulinowski KM, Monteiro-Riviere NA, Oberdörster G, Omenn GS, Pinkerton KE, Ramos KS, Rest KM, Sass JB, Silbergeld EK, Wong BA [2007]. Meeting report: hazard assessment for nanoparticles. Environ Health Perspect *115*(11):1654–1659.
- Schulte PA, Trout D, Zumwalde R, Kuempel E, Geraci CL, Castranova V, Mundt DJ, Mundt K, Halperin WE [2008]. Options for occupational health surveillance of workers potentially exposed to engineered nanoparticles: state of the science. J Occup Environ Med *50*:517–526.
- Papp T, Schiffmann D, Weiss D, Castranova V, Vallyathan V, Rahman Q [2008]. Human health implications of nanomaterial exposure. Nanotoxicol *2*:9–27.
- Ayres JG, Borm P, Cassee F, Castranova V, Donaldson K, Ghio A, Harrison RM, Hider R, Kelly F, Kooter I, Marano F, Maynard RL, Mud- way I, Nel A, Sioutas C, Smith S, Baeza-Squiban A, Cho A, Duggan S, Froines J [2008]. Evaluating the toxicity of airborne particulate matter and nanoparticles by measuring oxidative stress potential—a workshop report and consensus statement. Inhal Toxicol *20*:75–99.
- Castranova V [2008]. The Nanotoxicology Research Program in NIOSH. J Nanopart Res [www.springerlink.com/content/ c302m62567162 101/].
- Pacurari M, Yin XJ, Zhao J, Ding M, Leonard SS, Schwegler-Berry D, Ducatman BS, Sbarra D, Hoover MD, Castranova V, Vallyathan V [2008]. Raw single-walled carbon nanotubes induce oxidative stress and active MAPKs, AP-1, NFkB, and Akt in normal and malignant human mesothelial cells. Environ Health Perspect *116*:1211–1217.
- Pacurari M, Yin XJ, Ding M, Leonard SS, Schwegler-Berry D, Ducat- man BS, Chirila M, Endo M, Castranova V, Vallyathan V [2008]. Oxidative and molecular interactions of multi-walled carbon nano- tubes in normal and malignant human mesothelial cells. Nanotoxicol *2*:115–170.
- Vallyathan V, Pacurari M, Yin X, Zhao J, Ding M, Leonard S, Ducatman B, Castranova V [2008]. Alterations in molecular signaling of human mesothelial

cells exposed to raw single-walled carbon nanotubes. Respir Crit Care Med *177* Abs Issue:A219.

- Pacurari M, Robinson V, Castranova V, Leonard SS, Chen F, Vallyathan V, Bargern M [2008]. Does sandblasted metal attenuate or enhance the toxicity of freshly fractured silica? Toxicologist *1 02*:A300.
- Castranova V [2008]. Comparison of in vitro, pharyngeal aspiration, and inhalation results for single-walled carbon nanotubes. Toxicologist *1 02*:A2356.
- Hubbs AF, Mercer RR, Sriram K, Castranova V, Chen BT, McKinney W, Frazer DG, Batelli L, Willard P, Scabilloni J, Schwegler-Berry D, Porter D [2008]. Acute respiratory toxicolgic pathway of inhaled multi-walled carbon nanotubes. Vet Pathol *45*:A220.
- Schulte PA, Geraci CL, Zumwalde R, Hoover M, Castranova V, Kuempel E, Murashov V, Vainio H, Savolainen K [2008]. Sharpening the focus on occupational safety and health of nanotechnology. Scand J Work Environ Health, *34*(6):471–478.

Invited Presentations

- Castranova V [2007]. Nanotoxicology research in NIOSH. Toxicology forum, Washington, DC, January 30.
- Castranova V [2007]. Nanotechnology research in NIOSH. Meeting of the Northeastern chapter of the American Industrial Hygiene Association, Cleveland, OH, May 8.
- Castranova V [2007]. Nanotoxicology research in NIOSH. NASA Glenn Research Center, Cleveland, OH, May 8.
- Castranova V [2007]. Nanotoxicology research in NIOSH. American Industrial Hygiene Association conference and exposition, Philadelphia, PA, June 2–7.
- Castranova V [2007]. Nanotoxicology research in NIOSH. West Virginia University Cancer Center, NTRC meeting in Morgantown, WV, February 21.
- Castranova V [2007]. Critical toxicity parameters for nanoparticles vs. conventional particles. Presentation for the Department of Occupational Medicine, West Virginia University, Morgantown WV, August 28.
- Castranova V [2007]. Critical toxicity parameters for nanoparticles vs. conventional particles. Department of Physiology, West Virginia University, Morgantown, WV, August 30, 2007.
- Castranova V [2007]. What do we know about exposure and human health effects on nanoparticles. Annual scientific conference of the Occupational and Environmental Medical Association of Canada, Banff, Alberta, Canada, October 15 and 16.
- Castranova V [2007]. Critical toxicology parameters for nanoparticles vs. conventional particles. Conference on Bringing toxicology to global issues in occupational and environmental public health. Louisville, KY, October 18.
- Castranova V [2007]. Biological effects of nanoparticles. State-of-theArt Conference/ International Conference on Health Care Worker Health, Vancouver, Canada, October 26–28.

- Castranova V [2007]. NIOSH nanomaterials research activities. International Life Sciences Institute (ILSI) nanomaterials environmentalhealth and safety project committee, Washington, DC, November 8.
- Castranova V [2007]. Critical toxicity parameters for nanoparticles vs. conventional particles. Nanotechnology and occupational health and safety conference, Santa Barbara, CA, November 16 and 17.
- Castranova V [2007]. NIOSH nanotoxicology research. European NanOSH conference—nanotechnologies: a critical area in occupational safety and health, Helsinki, Finland, December 3–5.
- Castranova V [2008]. The Nanotoxicology Research Program in NIOSH. National Institute of Standards and Technology, Gaithersburg, MD, January 28.
- Castranova V [2008]. Nanotoxicology research in NIOSH: nanotechnology update for industrial hygienists, American Industrial Hygiene Association distance learning program January 31.
- Castranova V [2008]. Critical toxicity parameters for nanoparticles vs. conventional particles. Northern California Industrial Hygiene Society meeting, Palo Alto, CA, May 14.
- Castranova V [2008]. NIOSH nanotoxicology program. American Industrial Hygiene Association Conference and Exposition, Minneapolis, MN, May 31–June 5.
- Castranova V [2008]. NIOSH nanotoxicology research: biological activity of single-walled carbon nanotubes. American Industrial Hygiene Association conference and exposition, Minneapolis, MN, May 31–June 5.

Project 3. Systemic Microvascular Dysfunction: Effects of Ultrafine vs. Fine Particle

Principal Investigator Vincent Castranova, Ph.D.
Project Duration: FY 2006–2010

Critical Topic Area: toxicology and internal dose

Accomplishments and Research Findings

- Designed, constructed, and tested a generation system to produce a stable airborne concentration of fine and ultrafine TiO_2.
- Conducted inhalation exposures evaluating the dose-independent systemic microvascular response to pulmonary deposition of fine vs. ultrafine TiO_2 and found the following:
  - Inhalation of TiO_2 inhibited dilation of microvessels in the shoulder.
  - Inhalation of TiO_2 resulted in inhibition of the ability of microvessels in the shoulder to respond to dilators.
  - Systemic microvascular effects were demonstrated at lung burdens of TiO_2, which did not cause widespread pulmonary inflammation as judged by BAL markers.

- On an equivalent mass deposited basis, ultrafine TiO_2 was approximately 10-fold more potent in causing systemic microvascular dysfunction than fine TiO_2.
- Inhalation of TiO_2 inhibited nitric oxide production by systemic microvessels in response to dilators.
- The inhibition of nitric oxide production by microvessels is reversed by inhibitors of reactive oxygen species.
- The inhibition of microvessel dilation is partially reversed by blockade of neural input to the microvessel.
- The inhibition of microvessel dilation is partially reversed by systemic depletion of neutorphils.
- Inhalation of TiO_2 inhibited the ability of coronary vessel to dilate.

Publications and Abstracts

- Nurkiewicz TR, Porter DW, Hubbs AF, Stone S, Chen BT, Frazer D, Castranova V [2007]. Microvascular endothelium-dependent nitric oxide production is attenuated after inhalation of ultrafine particulate matter. FASEB J *21*:A846.
- Nurkiewicz TR, Porter DW, Hubbs AF, Cumpston JL, Chen BT, Frazer DG, Castranova V [2008]. Nanoparticle inhalation augments particle- dependent systemic microvascular dysfunction. Part Fibre Toxicol *5*(1).
- LeBlanc A, Hu Y, Muller-Delp J, Moseley A, Chen B, Frazer D, Castranova V, Nurkiewicz T [2008]. Particulate matter inhalation impairs coronary microvascular reactivity. Proc 25th Er Soc Microcir Conference, Budapest, Hungary, August 26–29, pp. 13–17.
- Nurkiewicz TR, Porter DW, Hubbs AF, Chen BT, Frazer DG, Castranova V [2008]. Ultrafine particulate matter inhalation attenuates micro- vascular endothelial nitric oxide production. Toxicologist *102*:A1255.

Invited Presentations

- Nurkiewicz TR, Porter DW, Hubbs AF, Millecchia LL, Donlin M, Chen TB, Frazer DG, Castranova V [2007]. Nanoparticle inhalation attenuates systemic microvascular endothelium-dependent nitric oxide production. World Congress for Microcirculation, Milwaukee, WI, August 15–19.
- Castranova V [2007]. Inhalation of ultrafine titanium dioxide augments particle-dependent microvascular dysfunction. Nanotoxicology conference, Venice, Italy, April 20.
- Castranova V [2007]. Cardiovascular effects of pulmonary exposure to nanoparticles. Toxicology and risk assessment conference, Cincinnati, OH, April 25.
- Castranova V [2007]. Inhalation of ultrafine titanium dioxide augments particle-dependent microvascular dysfunction. Center for Environmental Health, University of Montana, Missoula, MT, September 21.
- Castranova V [2008]. Systemic microvascular effects of pulmonary exposure to fine vs. ultrafine particles. University of Connecticut, Storrs, CT, April 7.

- Castranova V [2008]. Systemic microvascular effects of pulmonary exposure to fine vs. ultrafine particles. Meeting held at John Hopkins University, Baltimore, MD, June 10.
- Castranova V [2008]. Ambient particulates and/or nanoparticles: cardiovascular and pulmonary toxic and morphologic manifestations. International Conference on Particles: Risks and Opportunities, September 2–5.
- Chen BT, McKinney W, Schwegler-Berry D, Castranova V, Frazer D [2008]. Aerosolization and characterization of multi-walled carbon nanotubes for inhalation studies. American Association for Aerosol Research annual conference, Orlando, FL, October 20–24.
- Castranova V [2008]. Systemic microvascular effects of pulmonary exposure to particles. Presentation at the Harvard School of Public Health, Cambridge, MA, October 24.
- LeBlanc AF, Hu Y, Miller-Delp J, Chen D, Frazer BT, Castranova V, Nurkiewicz TR [2008]. Nanoparticle inhalation impairs endothelium- dependent vasoreactivity in coronary arterioles. American Heart Association conference, Washington DC, November 8–12.

Project 4. Particle Surface Area as a Dose Metric

Principal Investigator: Vincent Castranova, Ph.D.
Project Duration: FY 2004–2008

Critical Topic Area: toxicology and internal dose

Accomplishments and Research Findings

- Developed a method to improve dispersion of nanoparticles in biologically compatible suspension media.
- Determined that improved nanoparticle dispersion results in a greater inflammatory response after intratracheal instillation of ultrafine carbon black in rats.
- Compared ultrafine TiO_2 and carbon black to fine TiO_2 and carbon black, respectively, and determined that the ultrafine particles of both are more inflammogenic on a mass-dose basis.
- Determined that on an equivalent total particle surface area basis, the potencies of ultrafine versus fine TiO_2 or carbon black were not significantly different.
- Determined that pulmonary inflammation following intratracheal instillation of ultrafine TiO_2 was more persistent than after installation of ultrafine carbon black.
- Determined that in vitro treatment with ultrafine TiO_2 stimulates oxidant production and induction of molecular signaling events in macrophages.

Publications and Abstracts

- Shvedova A, Sager T, Murray A, Kisin E, Porter D, Leonard SS, Schwegler-Berry D, Robinson V, Castranova V [2007]. Critical issues in the evaluation of possible effects resulting from airborne nanoparticles. In: Monteiro-Riviere NA, Tran CL, eds. Nanotechnology: characterization, dosing, and health effects. New York: Informa Healthcare USA Inc., pp. 221–232.
- Sager T, Porter D, Robinson V, Lindsley WG, Schwegler-Berry DE, Castranova V [2007]. An improved method to disperse nanoparticles for in vitro and in vivo investigation of toxicity. Nanotoxicol *1*:118–129.
- Sager T, Robinson V, Porter D, Schwegler-Berry DE, Lindsley W, Castranova V [2007]. An improved method to prepare suspensions of nanoparticles for treatment of lung cells in culture or in vivo exposure by pharyngeal aspiration or intratracheal instillation. Toxicologist *96*:A1120.
- Kang JL, Moon C, Lee HS, Park E-M, Kim HS, Castranova V [2008]. Comparison of the biological activity between ultrafine and fine titanium dioxide particles in RAW 264.7 cells associated with oxidative stress. J Toxicol Environ Health *71* :478–485.
- Porter DW, Sriram K, Wolfarth MG, Jefferson AM, Schwegler-Berry DE, Andrew ME, Castranova V [2008]. A biocompatible medium for nanoparticle dispersion. Nanotoxicology *2*:144–154.
- Kang JL, Moon C, Lee HS, Lee HW, Park E-M, Kim HS, Castranova V [2008]. Comparison of the biological activity between ultrafine and fine TiO_2 particles in RAW 264.7 cells associated with oxidative stress. Toxicologist *1 02*:A1486.
- Sager T, Porter D, Castranova V [2008]. Pulmonary response to intratracheal instillation of fine or ultrafine carbon black or titanium dioxide: role of surface area. Toxicologist *1 02*:A 1491.
- Sager T, Kommineni C, Castranova V [2008]. Pulmonary response to intratracheal instillation of ultrafine versus fine titanium dioxide: role of surface area. Particle Fibre Toxicol *5*:17

Invited Presentations

- Castranova V [2008]. Intratracheal instillation of rats to ultrafine TiO_2 or ultrafine carbon black. Presentation at the Health and Environmental Sciences Institute, Washington, DC, May 8.

Project 5. Role of Carbon Nanotubes in Cardiovascular and COPD-Related Diseases

Principal Investigator: Petia Simeonova, Ph.D.
Project Duration: FY 2004–2008

Critical Topic Area: toxicology and internal dose

Accomplishments and Research Findings

- Evaluated cardiovascular effects resulting from respiratory exposure to purified SWCNT and MWCNT.
- A single exposure resulted in a treatment-dose dependent, distressed mitochondrial homeostasis represented by mitochondrial DNA damage, glutathione depletion, and increased protein carbonyl formation in aortic tissue. Oxidative alterations of mitochondria can trigger endothelial dysfunction, a leading mechanism in atherosclerosis progression.
- Multiple pulmonary exposures of apoE-/- mice, a model of human atherosclerosis, to SWCNT resulted in accelerated atherosclerosis.
- Respiratory exposure to SWCNT and MWCNT resulted in acute and chronic systemic responses related to potential cardiovascular adverse effects.
- Currently, there is no evidence for translocation of carbon nanotubes, in a measurable amount, from the lung into the systemic circulation. SWCNT and MWCNT deposited in the lung induced acute systemic inflammatory and prothrombotic responses. These effects are portrayed by a complex pattern of gene expression and protein blood analyses.
- Developed methodological approach used in the above studies for evaluating toxicity of respiratory exposure to particles. The approach will foster the development of predictive tests for estimating the toxicity of new nanomaterials based on their physiochemical characteristics and potential to induce oxidative stress, inflammation, and specific pulmonary and systemic toxicity.

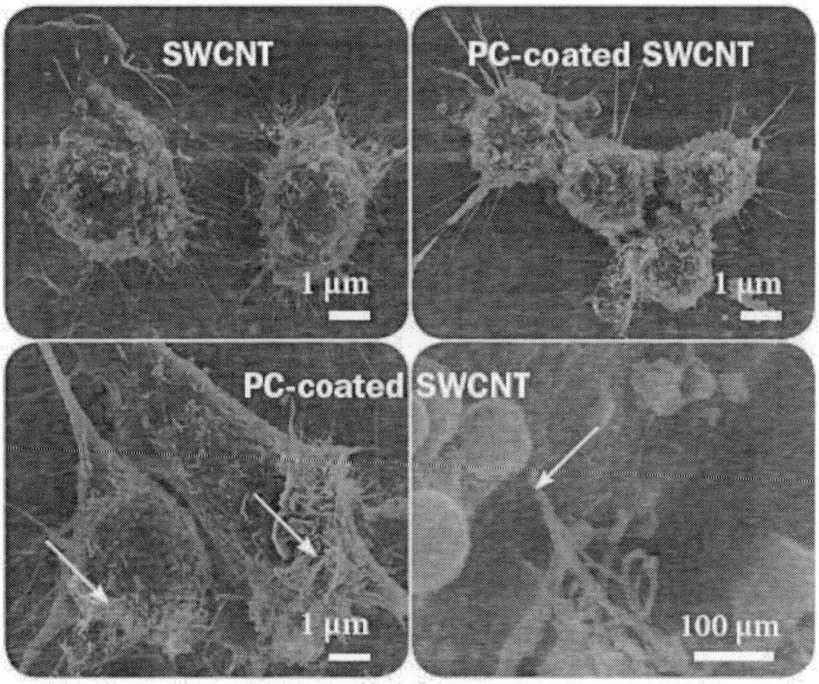

SEM of macrophages treated with PC- or PS-coated SWCNT

- Single exposure to CNTs is associated with a chronic and persistent release of osteopontin (OPN) into the systemic circulation. Preliminary data demonstrate a differential blood OPN response between female and male mice as a result of the CNT exposure. OPN may play a role in the CNT-induced local and cardiovascular responses.
- Evaluated cardiovascular effects of subchronic inhalation exposure to C60 particles (diameter 1 and 0.05 μm) in mouse and rat models (in collaboration with the National Institute of Environmental Health and Safety).
- Found that the initial screening for cardiovascular toxicity demonstrated changes in the expression of several genes related to stress response, including c-fos, and heat shock proteins 25, 70, and 90 in the aortas of the exposed mice. Heat shock proteins are incremental in triggering many autoimmune/chronic inflammatory diseases. In this regard, it has been suggested that there is a multifaceted role for heat shock proteins in atherosclerosis. Wild type mice (not prone to develop atherosclerosis) exposed to fullerenes developed cardiovascular stress responses, which may indicate a predisposition to atherogenesis.
- Discovered that endothelial cells exposed to CNTs resulted in a direct effect. The findings of this study demonstrated actin cytoskeleton disruption, VE-cadherin disorganization, reduced tubule formation, and concomitant diminished viability of human aortic endothelial cells as a result of exposure to purified SWCNT or MWCNT. These effects were dose dependent and most likely associated with mechanical damage of the cell membrane. The endothelial cells tolerate acute low concentrations of CNTs without changes in the viability, cytoskeleton, and function.
- Determined that study results demonstrate that the evaluation of systemic effects in parallel with pulmonary toxicity studies will provide a more complete toxicological assessment that can be used in predicting risk.

Publications and Abstracts

- Li Z, Hulderman T, Salmen R, Chapman R, Leonard SS, Shvedova A, Luster MI, Simeonova PP [2007]. Cardiovascular effects of pulmonary exposure to single-wall carbon nanotubes. Environ Health Perspect *1 15*(3):77–82. Nomimated for Charles C. Shepard Science Award, 2008, and was highlighted and discussed in the section "Hotoff-the-Presses Peer-Reviewed Research Article" of the EH&S Nano News [www.ehsnanonews.com/] as one of the first investigations on the potential cardiovascular effects of pulmonary exposure to SWCNT.
- Simeonava PP and Luster MI [2007]. In: Simeonova PP, Opopol N, Luster MI, eds. Nanotechnology: toxicological issues and environmental safety, NATO security through science series book. New York: Springer-Verlag, Springer-Verlag.
- NATO Security through Science Series Book *Nanotechnologytoxicological Issues and Environmental Safety* [2007]. Ed. Petia P. Simeonova, N. Opopol, MI Luster. Springer-Verlag.
- Simeonova PP [2007]. Nanoparticle exposure and cardiovascular effects—experimental data. In: Simeonova PP, Opopol N, Luster MI, eds. Nanotechnology:

toxicological issues and environmental safety, NATO security through science series book. New York: Springer-Verlag. pp. 53–65.

- Simeonova PP, Erdely A, Li Z [2007]. Carbon nanotube exposure and risk for cardiovascular effects. In: Monteiro-Riviere NA, Tran CL, eds. Nanotoxicology: charaterization, dosing, and health effects. New York: Informa Healthcare USA Inc., pp. 237–246.
- Simeonova PP, Li Z, Erdely A [2007]. Carbon nanotube respiratory exposure and risk from systemic effects. Toxicologist *96*(1): 137.
- Erdely A, Hulderman T, Salmen R, Liston A, Zeidler-Erdely PC, Simeonova PP [2008]. Carbon nanotube acute lung exposure induces plasminogen activator inhibitor 1. Toxicologist *102*(1): 151.
- Simeonova PP, Erdely A, Liston A, Roycroft J, Germolec D, Walker NJ, [abstract submitted]. Potential cardiovascular effects of Fullerene C60 inhalation exposure. Society of Toxicology annual meeting, Baltimore, Maryland, March 2009.
- Erdely A, Hulderman T, Salmen R, Liston A, Zeidler-Erdely PC, Simeonova PP [abstract submitted]. Time course of systemic effects following a single exposure to carbon nanotubes. Society of Toxicology annual meeting, Baltimore, Maryland, March 2009.
- Walker VG, Hulderman T, Simeonova PP [abstract submitted]. Comparison of carbon nanotube-induced cytotoxicity in A549 and normal human bronchial epithelial (NHBE) cells. Society of Toxicology annual meeting, Baltimore, Maryland, March 2009.
- Erdely A, Hulderman T, Salmen R, Liston A, Erdely P, Schwegler-Berry D, Castranova V, Koyama S, Kim Y-A, Endo M, Simeonova PP [2009]. Cross-talk between lung and systemic circulation during carbon nanotube respiratory exposure—potential biomarkers. Nano Lett *14*:9(1), pp. 36–43.
- Simeonova PP, Erdely A [in press]. Engineered nanoparticle respiratory exposure and potential risks for cardiovascular toxicity: predictive tests and biomarkers. Inhal Toxicol.
- Walker VG, Hulderman T, Schwegler-Berry D, Simeonova PP [in review]. Potential direct effects of carbon nanotubes on human aortic endothelial cells.

Invited Presentations

- Simeonova PP [2007]. Carbon nanotube respiratory exposure and risk from systemic effects. Society of Toxicology, Charlotte, NC, March, 25–29.
- Simeonova PP [2007]. Carbon nanotubes— toxicological evaluation. International meeting of nanotechnology, San Francisco, CA, November 5–7.
- Erdely A, Hulderman T, Salmen R, Liston A, Zeidler-Erdely PC, Simeonova PP [2008]. Crosstalk between the lung and blood following carbon nanotube exposure; potential biomarkers and implications for cardiovascular disease. Allegheny-Erie Chapter meeting, Society of Toxicology. Best overall poster award.
- Walker VG, Hulderman T, Simeonova PP [2008]. Direct exposure to carbon nanotube induces cellular changes in primary human endothelial cells. The American Society for Cell Biology annual meeting, San Francisco, CA, December 13–17.

- Simeonova PP [2008]. Engineered nanoparticle respiratory exposure and potential risk for cardiovascular toxicity. International inhalation symposium, Hanover, Germany, June 11–15.

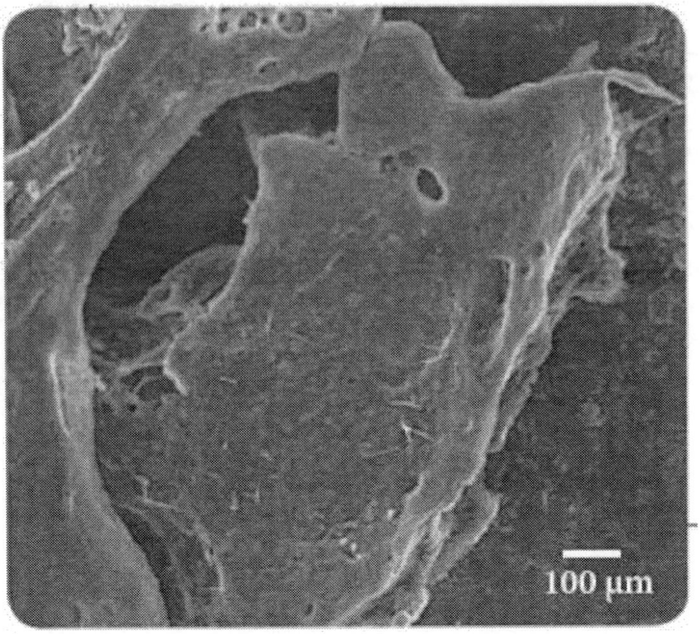

Scanning electron microscopy image of mouse macrophage with engulfed CNT

- Simeonova PP [2008]. Carbon nanotube pulmonary exposure— cardiovascular effects. American Association for the Advancement of Science annual meeting, Boston, MA, February 14–18.
- Simeonova PP [2008]. Air pollutants and atherosclerosis. Press briefing coordinated by the Director of the Office of Public Programs, American Association for the Advancement of Science, Boston, MA, February 14–18.
- Simeonova PP, Erdely A, Walker V [2008]. Reports on MWCNT systemic toxicity and discussions about collaborations with Dr. Endo and Japanese scientist delegation visiting NIOSH. Morgantown, WV, March 12.

Project 6. Investigations of Multi-Walled Carbon Nanotube Toxicity

Principal Investigator: Dale Porter, Ph.D.
Project Duration: FY 2007–2009

Critical Topic Area: toxicology and internal dose

Accomplishments and Research Findings

- Developed and validated a biocompatible dispersion medium for use in nanotoxicology studies.
- Investigated crystal structure of MWCNT using high resolution transmission electron microscopy.
- Determined surface chemistry of the MWCNT using x-ray photoelectron spectroscopy.
- Quantified metal contaminates and endotoxin levels preset in MWCNT.
- Evaluated dispersion status of hydrosol MWCNT using low resolution transmission electron microscopy.
- Determined zeta potential of hydrosol MWCNT.
- Determined the length and diameter of MWCNT.
- Conducted an in vivo dose-response and time-course study of MWCNT-induced toxicity after exposure by pharyngeal aspiration.
- Aspiration exposure study results indicate that MWCNT exposure induces dose- and time-dependent changes in pulmonary inflammation, damage, and fibrosis in the lung.
- Exposed mice by aspiration to gold-labeled MWCNT in order to determine MWCNT translocation. The samples have been collected and are currently being analyzed by neutron activation analyses.
- Built a MWCNT inhalation system in conjunction with the Developmental Engineering Research Team (DERT).
- Conducted a dose-response study to aerosolized MWCNT and found that inhaled MWCNT induces dose-dependent pulmonary inflammation and damage.

Publications and Abstracts

- Sager TM, Porter DW, Robinson VA, Lindsley WG, Schwegler-Berry DE, Castranova V [2007]. Improved method to disperse nanoparticles for in vitro and in vivo investigation of toxicity. Nanotoxicol *1*:118–129.
- Shvedova AA, Sager T, Murray AR, Kisin E, Porter DW, Leonard SS, Schwegler-Berry D, Robinson VA, Castranova V [2007]. Critical issues in the evaluation of possible adverse pulmonary effects resulting from airborne nanoparticles. In: Monteiro-Riviere NA, Tran CL, eds. Nanotoxicology: characterization, dosing, and health effects. New York: Informa Healthcare USA Inc., pp. 225–236.
- Sriram K, Porter DW, Tsuruoka S, Endo M, Jefferson AM, Wolfarth MG, Rogers GM, Castranova V, Luster MI [2007]. Neuroinflammatory responses following exposure to engineered nanomaterials. Toxicologist *96*:A1390.
- Sager T, Robinson V, Porter DW, Schwegler-Berry D, Lindsley W, Castranova V [2007]. An improved method to prepare suspensions of nanoparticles for treatment of lung cells in culture or in vivo exposure by pharyngeal aspiration or intratracheal instillation. Toxicologist *96*:A1 120.

- Porter DW, Sriram K, Wolfarth M, Jefferson A, Schwegler-Berry D, Andrew ME, Castranova V [2008]. A biocompatible medium for nanoparticle dispersion. Nanotoxicology *2*:144–154.
- Hubbs AF, Mercer RR, Sriram K, Castranova V, Chen BT, McKinney W, Frazer DG, Battelli L, Willard P, Scabilloni J, Schwegler-Berry D, Porter D [2008]. Acute respiratory toxicologic pathology of inhaled multi- walled carbon nanotubes. Vet Pathol *45*:786.
- Chen BT, McKinney W, Schwegler-Berry D, Castranova V, Frazer DG [2008]. Aerosolization and characterization of multi-walled carbon nanotubes for inhalation studies. American Association for Aerosol Research annual conference, Orlando FL, October 20–24.

Project 7. Pulmonary Toxicity of Metal Oxide Nanospheres and Nanowires

Principal Investigator: Dale Porter, Ph.D.
Project Duration: FY 2007–2010

Critical Topic Area: toxicology and internal dose

Accomplishments and Research Findings

- Developed and validated a biocompatible dispersion medium for use in nanotoxicology studies.
- Synthesized three engineered TiO_2 nanoparticles—nanospheres, short nanowires, and long nanowires (Nianqiang Wu, West Virginia University).
- Determined crystal structure of all three anatase TiO_2 nanoparticles.
- Characterized physical characteristics (diameter, length, width) of TiO_2 nanoparticles.
- Determined surface chemistry of the TiO_2 nanoparticles using electron spin resonance spectroscopy.
- Evaluated dispersion status of hydrosol TiO_2 nanoparticles using low resolution transmission electron microscopy and dynamic light scattering.
- Conducted an in vivo dose-response and time-course study of TiO_2 nanoparticles after exposure by pharyngeal aspiration. Results from animal aspiration study indicate that exposure to TiO_2 nanoparticles induces dose- and time-dependent changes in pulmonary inflammation and damage.
- Acted as co-principal investigator on National Science Foundation (NSF) grant, Correlation among physicochemical properties, photochemical fate, and toxicity of TiO_2 nanoparticles. Funding period September 1, 2008–August 31, 2011.

Publications and Abstracts

- Sager TM, Porter DW, Robinson VA, Lindsley WG, Schwegler-Berry DE, Castranova V [2007]. Improved method to disperse nanoparticles for in vitro and in vivo investigation of toxicity. Nanotoxicology *1*:118–129.
- Sager T, Robinson V, Porter DW, Schwegler-Berry D, Lindsley WG, Castranova V [2007]. An improved method to prepare suspensions of nanoparticles for treatment of lung cells in culture or in vivo exposure by pharyngeal aspiration or intratracheal instillation. Toxicologist *96*:A1 120.
- Porter DW, Sriram K, Wolfarth M, Jefferson A, Schwegler-Berry D, Andrew ME, Castranova V [2008]. A biocompatible medium for nanoparticle dispersion. Nanotoxicol *2*:144–154.
- Porter DW, Holian A, Sriram K, Wu N, Wolfarth M, Hamilton R, Buford M [2008]. Engineered titanium dioxide nanowire toxicity in vitro and in vivo. Toxicologist *102*:306.

Invited Presentations

- Hamilton R, Porter DW, Buford M, Sriram K, Wu N, Wolfarth M, Holian A [2008]. Engineered titanium dioxide nanoparticle bioactivity and toxicity are dependent on particle length and shape. Nanotoxicology— second international conference, Zurich, Switzerland. September 7–10.

Project 8. WC-Co Nanoparticles in Initiating Angiogenesis by Reactive Oxygen Species

Principal Investigator: Min Ding
Project Duration: FY 2009–2011. (Preliminary data gathering started in FY2008)

Critical Topic Area: toxicology and internal dose

Accomplishments and Research Findings

- This project will investigate the potential pulmonary carcinogenesis in response to WC-Co particles exposure using cell culture and animal model.
- Both fine and nanoparticles of WC-Co stimulate angiogenesis using the CEM assay.
- The mechanistic investigations (gene mutation, activation of transcription factors, reactive oxygen species [ROS] generation) indicate that WC-Co nanoparticles induce ROS production, which activates AKT and ERK signaling pathways in lung epithelial cells.
- ROS production results in transcriptional activation of AP- 1, NF- κB, and VEGF through AKT and ERK1/2 activation with the greater effect observed in nanoscale WC-Co when compared to fine-sized WC-Co at the same concentration.

Project 9. Evaluation of the Pulmonary Deposition and Translocation of Nanomaterials

Principal Investigator: Robert R. Mercer, Ph.D.
Project Duration: FY 2006–2010

Critical Topic Area: toxicology and internal dose

Accomplishments and Research Findings

- Found that aggregates of SWCNT disperse when deposited in the lungs.
- Demonstrated that diffuse interstitial fibrosis following exposure to SWCNT is due to a dispersed submicron form of SWCNT that had not yet been identified.
- Demonstrated that injury response (time course and resulting lesions) from the inhalation of SWCNT is similar to that produced by the aspiration of SWCNT, a less costly and more feasible delivery method.
- Received the 2008 Charles C. Shepard Science Award for outstanding scientific paper, *Alteration of Deposition Pattern and Pulmonary Response as a Result of Improved Dispersion of Aspirated Single-Walled Carbon Nanotubes in a Mouse Model.*
- Developed methods for labeling CNTs so that their potential translocation from the lungs to other organs can be studied.

Publications and Abstracts

- Mercer RR, Scabilloni JF, Wand L, Schwegler-Berry D, Shvedova AA, Castranova V [2007]. Dispersion significantly enhances the pulmonary toxicity of single-walled carbon nanotubes. Toxicologist *96*(231):A1115.
- Roberts JR, Mercer RR, Young S, Porter DW, Castranova V, Antonini JM [2007]. Inflammation and fate of quantum dots following pulmonary treatment of rats. Toxicologist *96*(230):A1111.

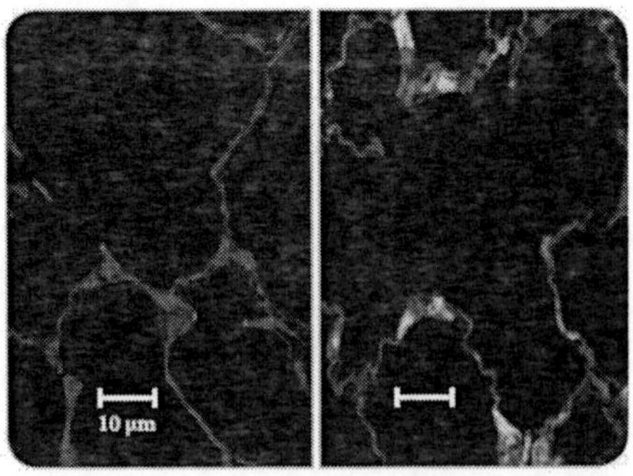

The image on the left is from a normal mouse lung while the one on the right if from a normal mouse lungs 30 days after exposure to SWCNT indicating an increase in collagen (stained bright yellow), the key element in fibrosis.

- Mercer RR, Scabilloni JF, Wang L, Battelli LA, Castranova V [2008]. Use of labeled single-walled carbon nanotubes to study acute translocation from the lungs. Toxicologist: *102*(S-1), A1399.
- Wang L, Castranova V, Rojanasakul Y, Lu Y, Scabilloni JF, Mercer RR [2008]. Direct fibrogenic effects of dispersed single-walled carbon nanotubes on human lung fibroblasts. Toxicologist: *102*(S-1), A1499.
- Wang L, Chanvorachote P, Toledo D, Stehlik C, Mercer RR, Castranova V, Rojanasakul Y [2008]. Hydrogen peroxide is a key mediator of Bcl-2 down regulation and apoptosis induction by cisplatin in human lung cancer cells. Mol Cancer Thera Mol Pharmacol *73*:119–127.
- Mercer RR, Scabilloni J, Wang L, Kisin E, Murray A, Schwegler-Berry D, Shvedova A, Castranova V [2008]. Alteration of deposition pattern and pulmonary response as a result of improved dispersion of aspirated single-walled carbon nanotubes in a mouse model. Am J Physiol Lung Cell Mol Physiol *294*:L87–97.
- Shvedova AA, Kisin E, Murray E, Johnson V, Gorelik O, Arepalli S, Hubbs AF, Mercer RR, Keohavong P, Sussman N, Jin J, Stone S, Chen BT, Deye G, Maynard A, Castranova V, Baron P, Kagan VE [2008]. Inhalation versus aspiration of single-walled carbon nanotubes in C57BL/6 mice: inflammation, fibrosis, oxidative stress, and mutagenesis. Am J Physiol Lung Cell Mol Physiol *295*:552–565.

Invited Presentations

- Mercer RR [2007]. Inhalation toxicology of SWCNT. Society for Risk Assessment Teleseminar, June 7.
- Mercer RR [2008]. Using labeled SWCNT to determine potential acute translocation from the lungs. Society of Toxicology annual meeting, March 16–20.

Project 10. Occupational Exposures and Potential Neurological Risks

Principal Investigator: Krishnan Sriram, Ph.D.
Project Duration: FY 2008–2012

Critical Topic Area: toxicology and internal dose

Accomplishments and Research Findings (Carbon nano-tubes study)

- Demonstrated the efficacy of a biocompatible dispersion medium for evaluating nanoparticle neurotoxicity.
- Evaluated neuroinflammatory responses following pulmonary and inhalation exposure to MWCNT.
- Evaluated the effects of MWCNT on blood-brain barrier integrity.
- Demonstrated regional heterogeneity in the blood-brain barrier changes and neuroinflammatory responses following pulmonary and inhalation exposure to MWCNT.

- Evaluated the effects of MWCNT in eliciting Alzheimer-like pathology.
- Demonstrated that pulmonary exposure to MWCNT up-regulates key genes involved in processing of the amyloid precursor protein, accumulation of which is a pathological hallmark of Alzheimer disease.

Accomplishments and Research Findings (welding fume study)

- Investigated the neurotoxicologic potential following pulmonary exposure to diverse welding fumes. Specifically, investigated the potential of welding fumes to cause dopaminergic neurotoxicity.
- Determined the regional metal distribution in the brain following pulmonary exposure to three types of welding fumes: gas metal arc-mild steel (GMA-MS), manual metal arc-hard surfacing (MMA-HS), and flux core arc-stainless steel (FCA-SS).
- Demonstrated the accumulation of manganese from welding fumes in target-dopaminergic brain areas.
- Demonstrated that with increasing solubility of welding fumes, agreater accumulation of manganese occurred in brain areas.
- Demonstrated that the potential to oxidize dopamine was directly proportional to the solubility of the welding fumes.
- Determined changes in T1 relaxation time (as an index of manganese accumulation) in specific brain areas using magnetic resonance imaging.
- Demonstrated that pulmonary exposure to GMA-MS, MMA-HS, and FCA-SS welding fumes caused loss of tyrosine hydroxylase protein, a marker of dopaminergic neurons.
- Demonstrated that acute inhalation exposure to GMA-MS elicited neuroinflammation and gliosis in specific brain areas, including dopaminergic targets.
- Demonstrated that acute inhalation exposure to GMA-MS alters theexpression of divalent metal transporters in distinct brain areas.

Accomplishments and Research Findings (titanium dioxide study)

- Investigated the influence of nanoparticle size and shape on neurotoxicologic potential, using nanosphere and nanowire TiO_2 particles.
- Evaluated neuroinflammatory responses following pulmonary exposure to TiO_2 nanosphere (NS), short TiO_2 nanowires (SNW) and long TiO_2 nanowires (LNW).
- Demonstrated that the potential of TiO_2 nanomaterials to elicit neuroinflammation was LNW>SNW>NS, indicating that particle size and shape influenced the toxicological response.

Publications and Abstracts

- Sriram K, Porter DW, Tsuruoka S, Endo M, Jefferson AM, Wolfarth MG, Rogers GM, Castranova V, Luster MI [2007]. Neuroinflammatory responses following exposure to engineered nanomaterials. Society of Toxicology annual meeting, Charlotte, NC, March 25–29, Toxicologist *96*(1):288.

- Antonini JM, Roberts JR, Sriram K, Benkovic SA, O'Callaghan JP, Miller DB [2008]. Extrapulmonary tissue distribution of metals following repeated lung exposures to welding fumes with different elemental profiles. Society of Toxicology annual meeting, Seattle, WA, March 16–20. Toxicologist *102*(1):226.
- Porter DW, Holian A, Sriram K, Wu N, Wolfarth M, Hamilton R, Buford M [2008]. Engineered titanium dioxide nanowire toxicity in vitro and in vivo. Society of Toxicology annual meeting, Seattle, WA, March 16–20. Toxicologist *102*(1):306.
- Porter D, Sriram K, Wolfarth M, Jefferson A, Schwegler-Berry D, Andrew ME, Castranova V [2008]. A biocompatible medium for nanoparticle dispersion. Nanotoxicology *2*(3): 144–154.
- Hubbs AF, Mercer RR, Sriram K, Castranova V, Chen BT, McKinney W, Frazer DG, Battelli L, Willard P, Scabilloni J, Schwegler-Berry D, Porter D [2008]. Acute respiratory toxicologic pathology of inhaled multi- walled carbon nanotubes. Vet Pathol *45*:786.

Project 11. Lung Effects of Resistance Spot Welding Using Adhesives

Principal Investigator: James Antonini, Ph.D.
Project Duration: FY 2007–2010

Critical Topic Area: toxicology and internal dose

Accomplishments and Research Findings

- Designed a resistance spot welding generator for animal inhalation studies.
- Nearly completed construction of the resistance spot welding generator.
- Characterized the physical and chemical properties of aerosols generated after heating adhesives used in resistance spot welding.

Publications and Abstracts

- Afshari AA, Antonini JM, Castranova V, Boylstein R, Kanwal R, Frazer DG [2008]. Design of an inhalation exposure system to study resistance spot welding fume characteristics and biological effects. Society of Toxicology annual meeting, Seattle, WA, March 16–20. Toxicologist *102*:226.

Invited Presentations

- Chen BT, Stone S, Schwegler-Berry D, Frazer A, Donlin M, Cumpston J, Afshari A, Frazer D, Castranova V, Antonini JM [2007]. Characterization of welding fume particles generated from a robotic welding system. American Association for Aerosol Research annual conference, Reno, NV, September 24.

- Antonini J [2008]. Health effects of welding. American Industrial Hygiene Association conference and exposition, Minneapolis, MN, May 31–June 5.

Project 12. Neurotoxicity after Pulmonary Exposure to Welding Fumes

Principal Investigator: James Antonini, Ph.D.
Project Duration: FY 2007–2010

Critical Topic Area: toxicology and internal dose

Accomplishments and Research Findings

- Characterized the physical properties of aerosol generated by welding and observed a significant number of welding particles in the nanoparticle-size range.
- Observed that, after inhalation, stainless steel particles persist in the lung and induce greater lung damage than mild steel particles.
- Observed the translocation of metals associated with welding fume (e.g., manganese and chromium) to other organ systems, including specific regions of the brain.
- Observed some alterations in inflammatory mediators in brain tissue from specific lung region.

Publications and Abstracts

- Santamaria AB, Cushing CA, Antonini JM, Finley BL, Mowat FS [2007]. Potential neurological effects of manganese exposure during welding: a state-of-the-science analysis. J Toxicol Environ Health, Part B, *10*:417–465.
- Antonini JM, Stone S, Roberts JR, Chen B, Schwegler-Berry D, Afshari AA, Frazer DG [2007]. Effect of short-term stainless steel welding fume inhalation exposure on lung inflammation, injury, and defense responses in rats. Toxicol Appl Pharmacol *223*:234–245.
- Antonini JM, Roberts JR [2007]. Comparison of lung injury and inflammation after repeated treatment with welding fumes collected from different welding processes. Society of Toxicology annual meeting, Charlotte, North Carolina, March 25–29. Toxicologist *96:*103–104.
- Antonini JM, Stone S, Roberts JR, Chen B, Frazer DG [2007]. Delayed inflammatory response after inhalation of stainless steel welding fume in rats. American Thoracic Society international conference, San Francisco, California, May 18–23. Am J Respir Crit Care Med *1 75*:A429.
- Zeidler-Erdely PC, Kashon ML, Battelli LA, Young S-H, Erdely AD, Roberts JR, Reynolds SH, Antonini JM [2008]. Lung inflammation and tumor induction in lung tumor susceptible A/J and resistant C57BL/6J mice exposed to welding fume. Part Fibre Toxicol *5*:12.
- Antonini JM, Roberts JR, Sriram K, Benkovic SA, O'Callaghan JP, Miller DB [2008]. Extrapulmonary tissue distribution of metals following repeated lung

exposures to welding fumes with different elemental profiles. Society of Toxicology annual meeting, Seattle, WA, March 16–20. Toxicologist *102*:226.

- Antonini JM, Stone S, Roberts JR, Schwegler-Berry D, Moseley A, Donlin M, Cumpston J, Afshari A, Frazer DG [2008]. Pulmonary effects and tissue distribution of metals after inhalation of mild steel welding fume. American Thoracic Society international conference, Toronto, Ontario, Canda, May 2008. Am J Respir Crit Care Med *177*:A9 10.

Invited Presentations

- Antonini J [2008]. Manganese exposure during welding. Institute of Occupational and Environmental Health, West Virginia University, Morgantown, WV, Occupational Medicine Grand Rounds Seminar, March 4.
- Antonini J [2008]. Health effects of welding. American Industrial Hygiene Association conference and exposition 2008, Minneapolis, MN, May 31–June 5.

Project 13. Potential Aneuploidy Following Exposure to Carbon Nanotubes

Principal Investigator: Linda Sargent, Ph.D.
Project Duration: FY 2007–2008

Critical Topic Area: toxicology and internal dose

Accomplishments and Research Findings

- Primary and immortalized human small airway epithelial cells were cultured and then exposed to SWCNTs or a positive control vanadium pentoxide to investigate the potential of CNTs to induce genetic damage in normal lung cells. The nanotubes had an average diameter of 1.1 nanometer and a length of 100–1,000 nanometers. Cellular tubulin, mitotic spindle integrity, and centriole number were determined by immunofluorescence for beta-tubulin and centrin and photographed using fluorescent and confocal laser scanning microscopy. The chromosome number was examined by fluorescent in situ hybridization.
- Binucleate cells, large polyploid appearing nuclei, bundled tubulin, and fragmented centrosomes were observed following 24 hours of exposure to either SWCNT or vanadium pentoxide.
- The SWCNT-treated cells were further examined by fluorescent in situ hybridization (FISH) using large genomic probes labeled with fluorescent tags. FISH showed that the large nuclei were polyploid. The number of chromosomes per cell was not normal (aneuploid). Mitotic spindles were found to have multiple poles and many cells had anaphase bridges with fragmenting chromatin. Using different interference confocal microscopy (DIC), SWCNT were observed to be inside the cells. DIC combined with fluorescent confocal laser imaging showed that many of the cells had SWCNT in the nucleus and within the cellular and mitotic tubulin. Some of the

binucleate cells appeared to be connected by SWCNT. CNTs in the mitotic spindle appear to distort the mitotic spindle apparatus. This damage was seen even at the lowest dose (24 μg/cm^3).

- The researcher contacted Dr. Jeffrey Salisbury to obtain a construct for one of the centrosome genes with a green fluorescent protein tag to allow confirmation of the centrosome fragmentation and also to follow the changes of the centrosome during the cell cycle. Dr. Salisbury will send antibodies for the centrosome that are more specific than commercially available antibodies. These antibodies can increase the sensitivity of the detection of the centrosome. Dr. Salisbury will also participate in imaging the SWCNT treated cells by electron microscopy both in the cell and following isolation of the centrosomes by a gradient. The centrosome work will be published separately.

Invited Presentations

- Sargent L [2008]. Genetic impact of carbon nanotubes and chemically specific genetic changes in lung adenocarcinoma. Naval Health Research Center, Detachment, Environmental Health Effects Laboratory, Wright-Patterson AFB, Dayton, OH, May 28.
- Sargent L [2008]. Genetic damage following carbon nanotube exposure. Mary Babb Randolph Cancer Center-National Institute of Occupational Safety and Health scientific retreat in cancer, February 12.

Project 14. Pulmonary Toxicity of Carbon Nanotube Particles

Principal Investigator: Anna Shvedova,
Ph.D. Project Duration: FY 2004–2008

Critical Topic Area: toxicology and internal dose

Accomplishments and Research Findings

- Low levels of antioxidants in vitamin E-deficient mice were associated with a higher sensitivity to SWCNT-induced acute inflammation, released LDH, total protein content, levels of proinflammatory cytokines, and enhanced profibrotic responses and collagen deposition.
- Exposure to SWCNTs markedly shifted the ratio of cleaved to full- length extracellular superoxide dismutase (EC-SOD) in the lungs. Found no differences in the EC-SOD responses to SWCNT-induced oxidative stress between vitamin E-deficient and vitamin E-sufficient animals.
- Sequential exposure to carbon nanotubes and bacteria enhanced pulmonary inflammation and infectivity.

- Decreased bacterial clearance in SWCNT-pre-exposed mice was associated with decreased phagocytosis of bacteria by macrophages and a decrease in nitric oxide production by these phagocytes.
- Failure of SWCNT-exposed mice to clear Lysteria led to a continued elevation in nearly all major chemokines and acute phase cytokines into the later course of infection.

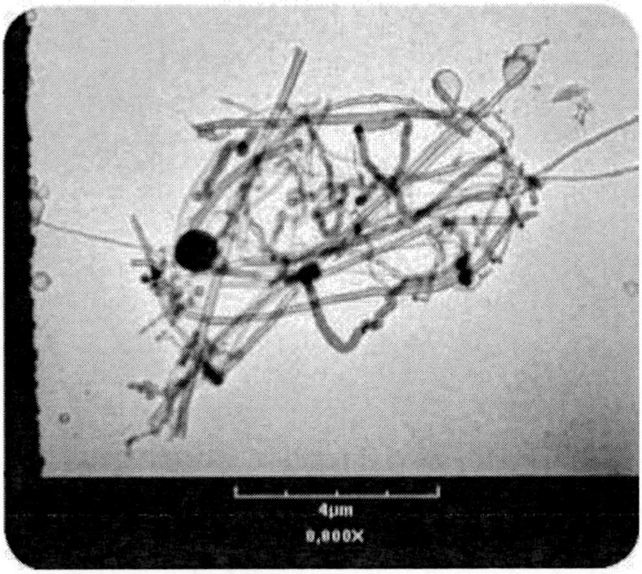

Electron micrograph of carbon nanotubes at 8,800x

- An increased accumulation of neutrophils and decreased fibrosis was observed in the lungs of NADPH oxidase-deficient C57BL/6 mice exposed to CNTs. This corresponded to elevated levels of apoptotic cells in the lungs, production of proinflammatory cytokines, decreased production of the anti-inflammatory and profibrotic cytokine, TGFbeta, and reduced levels of collagen deposition.
- NADPH oxidase is an important regulator of the transition from the acute inflammation to the chronic fibrotic stage in response to SWCNT.
- Developed a generation system that aerosolizes SWCNT for an inhalation study. The resulting aerosol was size-separated using a settling chamber and two cyclones to produce a respirable aerosol.
- The mass output efficiency of the entire system for producing a respirable aerosol from bulk material was estimated to be about 10%.
- Completed inhalation study of SWCNT in C57BL/6 mice.
- The inhalation of nonpurified SWCNT (iron content of 17.7% by weight) at 5 mg/m^3 5 hr/day for 4 days was compared with pharyngeal aspiration of varying doses (5–20 ug per mouse) of the same SWCNT.
- The chain of pathological events in both exposure routes was realized through synergized interactions of early inflammatory response and oxidative stress culminating in the development of multifocal granulomatous pneumonia and interstitial fibrosis.

- SWCNT inhalation was more effective than aspiration in causing inflammatory response, oxidative stress, collagen deposition, and fibrosis as well as mutations of K-ras gene locus in the lung of C57BL/6 mice.

Publications and Abstracts

- Shvedova AA, Sager T, Murray AR, Kisin E, Porter DW, Leonard SS, Schwegler-Berry D, Robinson VA, Castranova V [2007]. Critical issues in the evaluation of possible adverse pulmonary effects from airborne nanoparticles. In: Monteiro-Riviere NA, Tran CL, eds. Nanotoxicology: characterization, dosing, and health effects. New York: Informa Healthcare USA Inc.
- Kagan VE, Rao KMK, Kisin ER, Young S-H, Meighan T, Murray AR, Tyurina YY, Castranova V, Shvedova AA [2007]. Pulmonary effects of single-walled carbon nanotubes: inflammatory response, oxidative stress/signaling, and recognition by macrophages. In: Simeonova PP, Opopol N, Luster MI, eds. Nanotechnology: toxicological issues and environmental safety, NATO security through science series book, New York: Springer-Verlag, pp. 65–75.
- Shvedova AA, Kisin ER, Murray AR, Gorelik O, Arepalli S, Castranova V, Young SH, Gao F, Tyurina YY, Oury TD, Kagan VE [2007]. Vitamin- E deficiency enhances pulmonary inflammatory response and oxidative stress induced by single-walled carbon nanotubes in C57BL/6 mice. Toxicol Appl Pharmacol *221* (3):339–348.
- Fadeel B, Kagan V, Krug H, Shvedova A, Svartengren M, Tran L, Wiklund L [2007]. There's plenty of room at the forum: potential risks and safety assessment of engineered nanomaterials. Nanotoxicol *1* (2):73–84.
- Kisin ER, Murray AR, Keane MJ, Shi XC, Schwegler-Berry D, Gorelik O, Arepalli S, Castranova V, Wallace WE, Kagan VE, Shvedova AA [2007]. Single-walled carbon nanotubes: geno- and cytotoxic effects in lung fibroblast V79 cells. J Tox Environ Health, Part A, *70*(24):2071– 2079.
- Shvedova AA, Kisin ER, Murray AR, Gorelik O, Arepalli S, Castranova V, Young SH, Gao F, Tyurina YY, Oury T, Kagan VE [2007]. Vitamin-E deficiency enhances pulmonary inflammatory response and oxidative stress induced by single-walled carbon nanotubes in C57BL/6 mice. Toxicologist *96*(1):A179.
- Mercer RR, Scabilloni JF, Wang L, Schwegler-Berry D, Shvedova AA, Castranova V [2007]. Dispersion significantly enhances the pulmonary toxicity of single-walled carbon nanotubes. Toxicologist *96*(1) :A 1115.
- Baron PA, Deye GJ, Chen B-T, Schwegler-Berry D, Shvedova AA, Castranova V [2008]. Aerosolization of single-walled carbon nanotubes for an inhalation study. Inhal Tox *20*(8):751–760.
- Shvedova AA, Fabisiak JP, Kisin ER, Murray AR, Roberts JR, Tyurina YY, Antonini JM, Feng WH, Kommineni C, Reynolds J, Barchowsky A, Castranova V, Kagan VE [2008]. Sequential exposure to carbon nano- tubes and bacteria enhances pulmonary inflammation and infectivity. Am J Respir Cell Mol Biol *38*:579–590.
- Shvedova AA, Kisin E, Murray AR, Kommineni C, Castranova V, Fadeel B, Kagan VE [2008]. Accumulation of neutrophils and decreased fibrosis in NADPH oxidase-

deficient C57BL/6 mice exposed to carbon nanotubes. Tox Appl Pharm *231*(2):235–240.

- Shvedova AA, Kisin E, Murray AR, Johnson VJ, Gorerlik O, Arepalli S, Hubbs F, Mercer RR, Keohavong P, Sussman N, Jin J, Yin J, Stone S, Chen BT, Deye G, Maynard A, Castranova V, Baron PA, Kagan VE [2008]. Inhalation versus aspiration of single-walled carbon nanotubes in C57BL/6 mice: inflammation, fibrosis, oxidative stress, and mutagenesis. Am J Physiol Lung Cell Mol Physiol *295*(4):L552–565.
- Kisin E, Murray A, Keane M, Shi X, Schwegler-Berry D, Gorelik O, Arepalli S, Castranova V, Wallace W, Kagan V, Shvedova A [2008]. Single-walled carbon nanotubes: geno- and cytotoxic effects in lung fibroblast V79 cells. Toxicologist *102*(1):A441.
- Murray AR, Kisin E, Fabisiak JP, Roberts JR, Antonini JM, Kommineni C, Reynolds J, Barchowsky A, Castranova V, Kagan V, Shvedova AA [2008]. Combined exposure to carbon nanotubes and bacteria- enhanced pulmonary inflammation and infectivity. Toxicologist *102*(1):A1494.
- Shvedova AA, Kisin E, Murray AR, Johnson V, Gorelik O, Arepalli S, Hubbs AF, Mercer RR, Stone S, Frazer D, Chen T, Deye G, Maynard A, Baron P, Mason R, Kadiiska M, Stadler K, Mouithys-Mickalad A, Castranova V, Kagan VE [2008]. Inhalation of carbon nanotubes induces oxidative stress and cytokine response causing respiratory impairment and pulmonary fibrosis in mice. Toxicologist *102*(1):A1497.

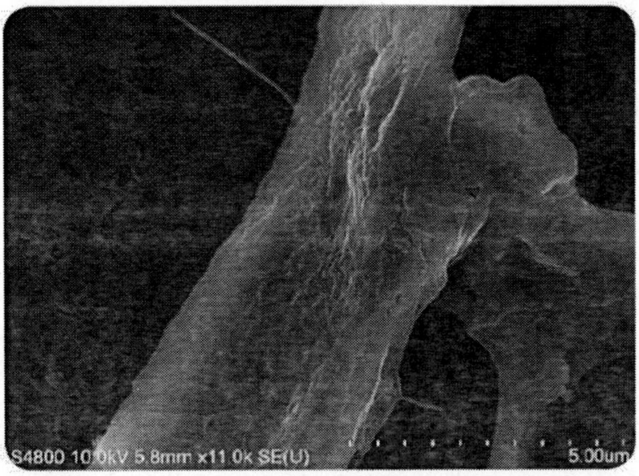

MWCNT in the lungs of a mouse 56 days after exposure. The shorter fiber on the right is penetrating the lung wall from the alveolar space. The longer fiber on the upper left has penetrated the outer wall of the lung into the pleural space.

Invited Presentations

- Shvedova AA [2007]. Pulmonary toxicity of singled-walled nanotubes. International conference on nanotoxicology, Venice, Italy, April 19 and 21.
- Shvedova AA [2007]. Pulmonary toxicity of nanoparticles. Experimental biology meeting, Washington, DC, April 30–May 3.

- Shvedova AA [2007]. Pulmonary toxicity of single-walled nanotubes in vivo: mechanisms and consequences. National Aeronautics and Space Administration/National Institute of Standards and Technology workshop on nanotube measurements. National Institute of Standards and Technology, Gaithersburg, MD, September 26–28.
- Shvedova AA [2007]. Inhalation exposure to single-walled carbon nanotubes. European NanOSH conference—nanotechnologies: a critical area in occupational safety and health, Helsinki, Finland, December 3–5.
- Shvedova AA [2008]. Single-walled carbon nanotubes: are they mutagenic? Mary Babb Randolph Cancer Center—National Institute of Occupational Safety and Health scientific retreat in cancer, Morgantown, WV, February 12.
- Shvedova AA, Kagan VE [2008]. Pulmonary toxicity of carbon nano- tubes in vivo: mechanisms, regulations, and reality. Nanomaterials: environmental risks and benefits and emerging consumer products, Faro, Portugal, April 27–30.
- Shvedova AA [2008]. Carbon nanotube in the rodent lungs: recognition and modulation of toxicity. American Thoracic Society, international conference, Toronto, Canada, May 16–2 1.
- Shvedova AA [2008]. Relevance of animal model studies: aspiration vs. inhalation. Second international conference: nanotoxicology, Zurich, Switzerland, September 7–10.

Project 15. Dermal Toxicity of Nanotube Particles

Principal Investigator: Anna Shvedova, Ph.D.
Project Duration: FY 2005–2010

Critical Topic Area: toxicology and internal dose

Accomplishments and Research Findings

- SWCNTs induced geno- and cytotoxic effects in lung fibroblast V79 cells when assessed using different test systems: the comet and micronucleus assays.
- Engineered human skin exposed to unpurified SWCNT showed increased epidermal thickness and accumulation and activation of dermal fibroblasts with increased collagen level and release of proinflammatory cytokines.
- Exposure of murine epithelial JB6 P+ cells to unpurified SWCNT (30% iron) resulted in the production of ESR detectable hydroxyl radicals and caused a significant dose-dependent activation of AP-1.
- Detected no significant changes in AP- 1 activation when partially purified SWCNT (0.23% iron) were introduced to JB6 P+ cells.
- NF-κB was activated in a dose-dependent fashion by exposure to both unpurified and partially purified SWCNT.
- Topical exposure of SKH- 1 mice to unpurified SWCNT induced free radical generation, oxidative stress, and inflammation, thus causing dermal toxicity.

Publications and Abstracts

- Fadeel B, Kagan V, Krug H, Shvedova A, Svartengren M, Tran L, Wiklund L [2007]. There's plenty of room at the forum: potential risks and safety assessment of engineered nanomaterials. Nanotoxicol *1* (2):73–84.
- Kisin ER, Murray AR, Keane MJ, Shi XC, Schwegler-Berry D, Gorelik O, Arepalli S, Castranova V, Wallace WE, Kagan VE, Shvedova AA [2007]. Single-walled carbon nanotubes: geno- and cytotoxic effects in fibroblast V79 cells. J Tox Environ Health, Part A, *70*(24):2071–2079.
- Tyurina YY, Basova LV, Konduru NV, Tyurin VA, Potapovich AI, Cai P, Bayir H, Stoyanovsky D, Pitt BR, Shvedova AA, Fadeel B, Kagan VE [2007]. Nitrosative stress inhibits the aminophospholipid translocase resulting in phosphatidylserine externalization and macrophage engulfment: implications for the resolution of inflammation. J Biol Chem *282*(11):8498–8509.
- Kagan VE, Tyurina YY, Konduru N, Basova LV, Stolz D, Kisin ER, Murray AR, Fadeel B, Shvedova AA [2007]. Targeted delivery of phosphatidylserine coated single-walled carbon nanotubes to macrophages. Toxicologist *96*(1):233.
- Feng W, Tyurina YY, Bayir H, Shvedova AA, Kagan VE [2007]. Phosphatidlyserine enhances recognition and uptake of single-walled carbon nanotubes by rat brain microglial cells. Toxicologist *96*(1):289.
- Murray AR, Kisin E, Kommineni C, Kagan VE, Castranova V, Shvedova AA [2007]. Single-walled carbon nanotubes include oxidative stress and inflammation in skin. Toxicologist *96*:A1406.
- Ding M, Zhao J, Bowman L, Leonard S, Lu Y, Kisin E, Murray A, Vallyathan V, Castranova V, Shvedova A [2008]. Induction of AP-1- MAPKs and NF-KB signal pathways by tungsten carbide-cobalt particles. Toxicologist *102*(1):A1031.
- Feng W, Konduru NV, Tyurina YY, Clark K, Stolz DB, Bayir H, Fadeel B, Shvedova AA, Kagan VE [2008]. Involvement of endocytosis in uptake of phosphatidylserine-coated single-walled carbon nanotubes by RAW 264.7 macrophages. Toxicologist *102*(1):213.

Invited Presentations

- Shvedova AA [2008]. Single-walled carbon nanotubes: are they mutagenic? Mary Babb Randolph Cancer Center—National Institute of Occupational Safety and Health scientific retreat in cancer, Morgantown, WV, February 12.
- Murray AR, Kisin ER, Leonard SS, Kommineni C, Kagan VE, Castranova V, Shvedova AA [2008]. Induction of oxidative stress and inflammation in the skin following exposure to single-walled carbon nano- tubes. Zurich, Switzerland, March 12–15.

Project 16. Specific Biomarkers for Unusual Toxicity of Nanomaterials

Principal Investigator: Liying Wang Rojanasakul, Ph.D.
Project Duration: FY 2007–2008

Critical Topic Area: toxicology and internal dose

Accomplishments and Research Findings

- In vitro cellular models were developed for fibrogenic and biomarker studies of CNTs, such as collagen and collagenase production from fibroblasts and TGF-β production from lung epithelial cells.
- Data from this project provided important preliminary findings used in obtaining NORA grant funding.

Publications and Abstracts

- Wang L, Castranova V, Rojanasakul Y, Lu Y, Scabilloni J, Mercer RR [2008]. Direct fibrogenic effects of dispersed single-walled carbon nanotubes on human lung fibroblasts. Society of Toxicology annual meeting, Washington, DC, March 16–20. Toxicologist *102*:A1499.
- Wang L, Castranova V, Mercer RR, Wu N, Tiao Li T, Li S, Hall J, Li M, Rojanasakul Y [abstract accepted]. Nanoparticle dispersion method using natural lung surfactant. Society of Toxicology annual meeting, Baltimore, Maryland, March 2009.

Invited Presentations

- Wang L [2007]. Carbon nanotubes exhibit tumorigenic and fibrogenic potential in vitro. West Virginia University research meeting, Morgantown, WV, September.
- Wang L [2008]. Meeting with the National Institute of Occupational Health and Poison Control on nanotoxicology research, Beijing, China, November 16 and 17.
- Castranova V [2008]. The ability of in vitro tests to predict pulmonary responses to SWCNT. John Hopkins University, Baltimore, MD, June 11.

Project 17. Pulmonary Toxicity of Diesel Exhaust Particles

Principal Investigator: Jane Ma, Ph.D.
Project Duration: FY 2003–2008

Critical Topic Area: toxicology and internal dose

Accomplishments and Research Findings

- Diesel-exhaust-particle (DEP) exposure induces lung inflammation and injury, which are mediated through reactive oxygen/nitrogen species (ROS/RNS) generation by alveolar macrophages (AM).
- C57B/6J wild type (WT) and iNOS knockout (iNOS KO) mice were exposed to saline, carbon black (35 mg/kg), or organic extract of DEP (DEPE) by aspiration and sacrificed at 1, 3, and 7 days postexposure. Carbon black, but not DEPE, significantly induced neutrophil infiltration in both WT and iNOS KO mice.
- Carbon black, but not DEPE, exposure results in inflammatory lung injury, suggesting that DEP-induced pulmonary inflammation is derived from the particulate. Nitric oxide (NO) does not play a significant role in Carbon-black-induced polymorphonuclear neutrophils (PMN) infiltration, air/capillary damage, and cytotoxicity.
- Carbon-black-exposed AM exhibited enhanced superoxide anion generation, peaking at 3 days post exposure. Carbon black exposure also induced mitochondrial damage, monitored as reduction of mitochondrial membrane potential in AM, which was not markedly affected by the NO deficiency. In contrast, DEPE did not significantly induced ROS generation or mitochondrial function in AM.
- Carbon black and DEPE did not significantly affect adenosine triphosphate (ATP) levels in either WT or iNOS KO mice, because mitochondrial oxidative phosphorylation pathway is not the major energy source for AM.
- Transmission electron microscopy analysis has shown that animals exposed to DEP or carbon black, but not DEPE, were heavily laden with particle-filled vacuoles in the lung.
- Vacuoles containing DEP from DEP-exposed animals were frequently associated with mitochondria or were in direct membrane-to-membrane contact with mitochondria, demonstrating that mitochondrial function was compromised. In contrast, carbon-black-filled vacuoles did not display functionally impaired mitochondria, and no intracellular vacuoles were found in animals exposed to DEPE.
- DEP and carbon black both induced DNA damage in AM using comet assay. DEPE did not.

Publications and Abstracts

- Ma JYC, Millecchia L, Barger MW, Ma JKH, Castranova V [2007]. Diesel exhaust particle-induced oxidant injury and cellular responses in wild type and inducible nitric oxide synthase-deficient (iNOS KO) mice: roles of particle core and adsorbed organics. Toxicologist *96*(S- 1):A183.
- Yin XJ, Dong CC, JYC, Roberts JR, Antoni J, Ma JKH [2007]. Suppression of phagocytic and bactericidal functions of rat alveolar macrophages by the organic component of diesel exhaust particles. J Toxicol Environ Health, Part A, *70*(10):820–828.

Invited Presentations

- Ma JYC [2008]. Effects of exposure to diesel exhaust particles on the susceptibility of the lung to infection. The Feinstein Institute for Medical Research, Manhasset, NY, April 8.

Project 18. Induction of Lung Fibrosis by Cerium Oxide in Diesel Exhaust

Principal Investigator: Jane Ma, Ph.D.
Project Duration: FY 2009–2012 (preliminary data gathering started in FY 2008)

Critical Topic Area: toxicology and internal dose

Accomplishments and Research Findings

I. In vitro studies
- Cerium oxide did not induce intracellular oxidant generation or alter mitochondrial function of alveolar macrophages (AM) but significantly damaged AM membrane integrity, induced apoptosis, affected macrophage inflammatory responses, including increased TNF-α, but reduced IL- 12 production. These results suggest that cerium oxide can cause direct cellular damage to AM.
- Performed tests suggest that cerium oxide, under coexposure, may elicit new cellular toxicity in addition to DEP-induced toxicity and that cerium oxide suppresses cellular defense capability.
- Previous studies have shown that nitric oxide plays an important role in the DEP-induced IL-12 production by AM. The decreased nitric oxide production in response to LPS challenge by cerium oxide is probably due to its redox characteristic, subsequently leading to the suppressed IL- 12. In the coexposure cell culture, cerium oxide depressed DEP- and LPS-induced IL-12 secretion, suggesting that cerium oxide suppresses cellular defense capability.

II. In vivo studies
- Cerium oxide administered intratracheally to Sprague Dawley rats induced significant neutrophil infiltration and elevated lactate dehydrogenase activity and albumin content in the BAL fluid, suggesting that these particles induced inflammation, cytotoxicity, and epithelial damage.
- Found a direct interaction of cerium oxide with AM, resulting in elevated AM phagocytic activity as monitored using confocal microscopy.
- Cerium oxide exposure increased reactive oxygen species production in response to zymosan challenge as indicated by enhanced chemiluminescence generation.
- Cerium oxide significantly reduced nitric oxide production by AM in response to ex vivo LPS challenge, demonstrating that these particles may modify AM host defense capability.

Publications and Abstracts

- Ma JY, Hongwen Z, Barger M, Castranova V, Ma JK [2008]. Effects of cerium oxide on rat primary alveolar macrophages. Toxicologist *102*(1):A1513.
- Ma JY, Zhao H, Barger M, Murali R, Meighan T, Castranova V, Ma JK [abstract accepted]. Pulmonary responses to diesel fuel catalyst cerium oxide nanoparticles. Society of Toxicology annual meeting, Baltimore, MD, March 2009.

Project 19. Potential Effects of Silicon-based Nanowires on Lung Toxicity

Principle Investigators: Stephen S. Leonard and Jenny R. Roberts
Project Duration: FY 2008–2010

Critical Topic Area: toxicology and internal dose

Accomplishments and Research Findings

- Completed physical characterization of silicon-based nanowires generated by IBM and titanium nanowires to be used as a control/reference material.
- Characterized and quantified nanowire-associated reactive oxygen species (ROS) using a cell-free model system and quantified nanowireinduced free radical generation in vitro in alveolar macrophages.
- Measured and evaluated pulmonary toxicity parameters in vivo using a titanium nanowire to assess the potential of the material as a control. The nanowires induced a transient, dose-dependent response in the lungs of rats that included increases in indicators of inflammation and oxidative stress.
- Evaluated the ability of pulmonary exposure to titanium nanowires to alter responses to a pulmonary infection. Only the highest dose of nanowires altered response to infection, inducing an increase in resolution of infection.
- Evaluated a potential fluorescent marker, quantum dots, for tracking distribution of nanomaterials in the lung.

Publications and Abstracts

- Roberts JR, Mercer RR, Young S-H, Porter DW, Castranova V, Antonini JM [2007]. Inflammation and fate of quantum dots following pulmonary treatment of rats. Society of Toxicology annual meeting, Charlotte, North Carolina, March 25–29. Toxicologist *96*:230.
- Leonard SS, Castranova V, Chen BT, Schwegler-Berry D, Hoover M, Piacitelli C, Gaughan D [2007]. Particle-size-dependent radical generation from wildland fire smoke. Toxicol *236*:103–113.
- Pacurari M, Yin XJ, Zhao J, Ding M, Leonard SS, Schwegler-Berry D, Ducatman BS, Sbarra D, Hoover MD, Castranova V, Vallyathan V [2008]. Raw single-wall carbon nanotubes induce oxidative stress and activate MAPKs, AP-1, NF-kB, and

Akt in normal and malignant human mesothelial cells. Environ Health Perspect *116*(9):1211–1217.

- Roberts JR, Schwegler-Berry D, Leonard SS, Karim A, Tirumala V, Antonini JM, Castranova V [2008]. Pulmonary toxicity associated with nondispersed titanium dioxide nanorods. Society of Toxicology annual meeting, Seattle, Washington, March 16–20. Toxicologist *102*:308.
- Ding M, Zhao J, Bowman LL, Leonard SS, Lu Y, Vallyathan V, Castranova V, Shvedova AA [2008]. Induction of AP-1-MAPKs and NF-kB signal pathways by tungsten carbide-cobalt particles. Toxicologist *102*:2 12, A1031.
- Roberts JR, Antonini JM, Porter DW, Castranova V, Mercer RR [2008]. Characterization of pulmonary responses following treatment of rats with fluorescently labeled quantum dots with different surface functional groups. American Thoracic Society international conference, Toronto, Ontario, Canada, May 16–17, 2008. Am J Respir Crit Care Med *117*:A49.
- Roberts JR, Schwegler-Berry D, Chapman R, Antonini JM, Scabilloni JF, Castranova V, Mercer RR [abstract submitted]. Biodistribution of quantum dots after pulmonary exposure in rats. Society of Toxicology annual meeting, Baltimore, MD, March 2009.
- Chapman R, Roberts JR, Castranova V, Leonard SS [abstract accepted]. Generation of reactive oxygen species by silicon nanowires. Society of Toxicology annual meeting, Baltimore, MD, March 2009.
- Rushton EK, Jiang J, Leonard SS, Eberly S, Castranova V, Biswas P, Elder A, Gelein R, Finkelstein J, Oberdorster G [accepted for publication]. Predicting nanoparticle toxicity: evaluation of screening assays. Environ Health Perspect.

Project 20. Cell-Based Assessment for Iron Nanoparticle-Induced Health Risks

Principal Investigator: Yong Qian
Project Duration: FY 2008–2010

Critical Topic Area: toxicology and internal dose

Accomplishments and Research Findings

- Developed a cell-based model for assessing nanoparticle particle- induced health risks.
- Trained one post-doctoral fellow and one graduate student in nano- technology research for advancement of their professional careers in the field.
- Data from this project provided key preliminary results for a successful Pilot Funding Grant from the Mary Babb Randolph Cancer Center in the Health Sciences Center at West Virginia University.

Publications and Abstracts

- Apopa PL, Qian Y, Guo NL, Schwegler-Berry D, Pacurari M, Porter D, Shi X, Vallyathan V, Castranova V, Flynn DC [in press]. Iron oxide nanoparticles induce human microvascular endothelial cell permeability through reactive oxygen species production and microtubule remodeling. Particle and Fibre Toxicology.
- Apopa PL, Guo NL, Schwegler-Berry D, Pacurari M, Porter D, Shi X, Vallyathan V, Castranova V, Flynn DC, Qian Y [in press]. Iron oxide nanoparticles induce human microvascular endothelial cell permeability through reactive oxygen species production and microtubule remodeling. Annual Society of Toxicology Meeting.

Project 21. Assessment of Carbonaceous Materials on Mutagenicity

Principal Investigator: Anna A. Shvedova, Ph.D., D.Sc.
Project Duration: FY 10/1/08–9/30/09

Critical Topic Area: toxicology and internal dose

Accomplishments and Research Findings

- Inhalation exposure of C57BL/6 mice (5 mg/m^3, 5 hr a day, 4 days) to respirable SWCNT caused formation of anuclear macrophages with fibrillar cytoplasm along with mitotic changes was seen in the lung interstitial cells. Pattern of anaphase bridges in macrophages was clearly detected at 7 and 28 days post inhalation. Therefore, such mitotic changes within macrophage with anaphase bridges indicated feasible spindle aberrations.
- Post-exposure findings (7-days) revealed observable foci of bronchiolar epithelial cell hypertrophy in all exposed mice. Post-exposure findings (28-days) revealed bronchiolar epithelial cell hypertrophy and hyperplasia in all exposed mice.
- Severe oxidative stress observed in the lung of mice exposed to inhalable SWCNT revealed depletion of GSH, protein thiol oxidation, accumulation of lipid peroxidation products, reduced total antioxidant reserves, and augmentation of 8-hydroxy-2'-deoxyguanosine (8-OHdG)—a marker of oxidative D damage.
- K-ras mutations were observed in the lung of C57BL/6 mice after inhalation exposure to SWCNT. The first 2 mutations in DNA samples corresponded to a change in the wild type K-ras gene codon 12 (GGT, glycine) to AGT (lane 2, serine) and GAT (lane 3, aspartate). The third mutation was a double mutation consisting of a GGT to GAT (glycine to aspartate at codon 12) and a GTG to ATG (valine to methionine at codon 8)].
- One of the mutations found in SWCNT exposed mice by inhalation route (at day 28 post exposure) consisted of a double mutation occurring at codons 12 and 8 [GGT to GAT and GTG to ATG (valine to methionine), respectively. The role of this double mutation is unknown and may be specific for SWCNT exposure.

Publications and Abstracts

- Shvedova AA, Kisin E, Murray AR, Johnson, VJ, Gorerlik O, Arepalli S, Hubbs F, Mercer RR, Keohavong P, Sussman N, Jin J, Yin J, Stone S, Chen BT, Deye G, Maynard A, Castranova V, Baron PA, Kagan VE [2008]. Inhalation versus aspiration of single walled carbon nanotubes in C57BL/6 mice: inflammation, fibrosis, oxidative stress and mutagenesis. Am J Physiol Lung Cell Mol Physi *295*(4):L552–65.
- Elder A, Lynch I, Grieger K, Chan-Remillard S, Gatti A, Gnewuch H, Kenawy E-R, Korensten R, Kuhlbusch T, Linker F, Matias S, monteiroRiviere N, pinto V, Rudnitsky R, Savolainen K, Shvedova A [2008]. Deposition of nanoparticles as a function of their interactions with biomolecules. In "Nanomaterials Risks and Benefits", Eds. Linkov & Steevens, Springer, The Netherland, pp. 3–29, (Proceedings of the NATO Advanced Research Workshop on Nanomaterials: Environmental Risks and Benefits Faro, Portugal, April 27–30.
- Shvedova AA, Kisin ER, Porter D, Schulte P, Kagan VE, Fadeel B, Castranova V [2008]. Mechanisms of pulmonary toxicity and medical applications of carbon nanotubes: Two faces of Janus? Pharmacol Ther December 6, PMID:19103221. [Epub ahead of print].

Invited Presentation

- Shvedova AA [2008]. Single-walled carbon nanotubes: Are they mutagenic? Mary Babb Randolph Cancer Center. National Institute of Occupational Safety and Health MBRCC, NIOSH Scientific Retreat in Cancer, Morgantown, WV, February 12.

Project 22. Generation and Characterization of Nanoparticles

Principal Investigator: Bon-Ki Ku, Ph.D.
Project Duration: FY 2004–2008

Critical Topic Area: measurement methods

Accomplishments and Research Findings

- Developed an electrospray generation system to produce SWCNT aerosols with well-defined morphology for the purpose of evaluation instrument response to different morphology and possibly applying them to in vitro cell culture toxicity tests.
- Field-tested size-selected nanoparticle measurement techniques and sampling methods in workplaces for physical and chemical analysis using transmission electron microscopy.
- Characterized purified SWCNTs generated by electrospraying of suspensions.
- Conducted research on the relationship among electrical mobility, mass, and size for nanodrops 1–6.5 nm in diameter.

- Demonstrated that the established relation (which is used in the scanning mobility particle sizer) between electrical mobility and mass diameter for a sphere can be applied with confidence over the whole size range down to 1.3 nm.
- Established significant partnerships with the University of Minnesota, University of Iowa, and Yale University.
- Investigated the relationship between aerodynamic and mobility diameters of SWCNTs and MWCNTs and airborne carbon nanofiber.
- In conjunction with the University of Minnesota, applied a new method to characterize the structure of airborne SWCNTs. The results obtained are expected to be used by toxicologists and other scientists to investigate and identify physical properties that may contribute to observed particle toxicity.
- Observed anomalous behavior in state-of-the-art aerosol instrumentation. This research is expected to lead to the development of effective monitoring methods and new/improved instrument designs that are needed for accurate characterization of worker exposures.
- In conjunction with Yale University, contributed a large amount of data on mobility and mass relations for a significant number of different materials in the size range of a few nanometers. The data are valuable in giving a basis for accurate measurements of nanoparticles at very small sizes and are expected to be used in future research.
- Characterized bipolar charging characteristics of carbon nanofibers to better understand particle charging-based instrument response to different particle morphologies, such as spherical and fiber shape.

Publications and Abstracts

- Maynard AD, Ku B-K, Emery M, Stolzenburg M, McMurry P [2007]. Airborne single-walled carbon nanotube agglomerates show particle size-dependent physicochemical structure. J Nanopart Res *9*:85–92.
- Ku B-K, Maynard AD, Baron PA, Deye G [2007]. Observation and measurement of anomalous responses in a differential mobility analyzer by ultrafine fibrous carbon aerosols. J Electrostat *65*:542–548.
- Ku B-K [2008]. Nanomaterials and occupational health. Indust Health *2*:44–49 (in Korean).
- Ku B-K, Kulkarni P [in press]. Morphology of single-wall carbon nanotube aggregates generated by electro spray of aqueous suspensions. J Nanopart Res.
- Ku B-K, Fernandez de la Mora J [2009]. Relation between electrical mobility, mass, and size for nanodrops 1–6.5 nm in diameter in air. Aerosol Sci Technol *43*(3), 241–249.
- Shvedova AA, Kisin E, Murray AR, Johnson V, Gorelik O, Arepalli S, Hubbs AF, Mercer RR, Keohavong P, Sussman N, Jin J, Yin J, Stone S, Chen BT, Deye G, Maynard A, Castranova V, Baron PA, Kagan VE [in press]. Inhalation versus aspiration of single-walled carbon nanotubes in C57BL/6 mice: inflammation, fibrosis, oxidative stress, and mutagenesis. Am J Physiol Lung Cell Mol Physiol.

Invited Presentations

- Ku B-K [2007]. Characterization of purified single-walled carbon nanotube aerosols generated by electro spraying of suspension. International symposium on nanotechnology, occupational, and environmental health, Taipei, Taiwan, August 29–September 1.
- Ku B-K, Fernandez de la Mora J, Ude S [2007]. Relation between electrical mobility, mass, and size in the nanometer range of charged nanoparticles generated by electrosprays. American Association for Aerosol Research annual conference, Reno, NV, September 24–28.
- Ku B-K [2007]. Nanoparticles and occupational health: toward safe nanotechnology conference. University of Kentucky, Lexington, KY, April 24.
- Ku B-K [2007]. Nanoparticles, occupational and environmental health: physical characterization of nanoparticles conference. Korea Institute of Machinery and Materials, Daej eon, Korea, September 3 and 4.
- Ku B-K [2007]. Contemporary issues in occupational health: nanoparticles and occupational health conference. University of Yonsei, Seoul, Korea, September 5.
- Ku B-K, Fernandez de la Mora J [2008]. Electrical mobility, mass, and size for nanodrops 1–3.5 nm in diameter. American Association for Aerosol Research annual conference, Orlando, FL, October 20–24.
- Kulkarni P, Deye GJ, Ku B-K, Baron PA [2008]. Relationship between aerodynamic and mobility diameters of single- and multi-walled carbon nanotube aerosols. American Association for Aerosol Research annual conference, Orlando, FL, October 20–24.

Project 23. Dustiness of Nanomaterials

Principle Investigator: Douglas E. Evans, Ph.D.
Project Duration: FY 2007–2009

Critical Topic Areas: measurement methods, exposure assessment, fire and explosion safety

Accomplishments and Research Findings

- Completed pilot phase of the research.
- Evaluated novel method for testing powders and found it to be suitable for testing nanomaterials. Advantages of the method include
  - Small powder samples required (mg quantities), allowing for a potentially greater number of materials to be tested
  - Complete enclosure of the test system so that the operator is not exposed to potentially toxic materials
  - Relevance to the workplace

- Tested approximately 20 fine and nanoscale materials—consisting of at least three replicate tests—for total and respirable mass. Values span 2 orders of magnitude.
- Modified the testing system and protocol to investigate humidity effects on powder dispersal.
- Designed a new dispersal chamber, allowing for a greater variety of measurements to be made.
- In February 2007, visited the UK Health and Safety Laboratory in Buxton.
    - Discussed issues related to workplace measurement of nanomaterials and dustiness testing methods.
    - Conducted dustiness tests of the novel method in parallel with the more traditional (UK) rotating-drum method using the same materials. Technical issues were identified with both methods.
- In early 2008, submitted a full NORA/NTRC proposal for the FY 2009 solicitation to support continuing dustiness research.

Invited Presentations

- Evans DE [2007]. Working at the nanoscale: the occupational health and safety perspective. Health and Safety Laboratory, Buxton, United Kingdom, February 20.
- Mark D, Bard D, Wake D, Hoover MD, Stefaniak AB, Methner MM, Evans DE, Geraci CL [2008]. Problems and experience of measuring worker exposure to engineered nanoparticles. International inhalation symposium: benefits and risks of inhaled engineered nanoparticles, Hanover, Germany, June 11–14

Project 24. Nanoaerosol Monitoring Methods

Principle Investigators: M. Eileen Birch, Ph.D. and Douglas E. Evans, Ph.D.
Project Duration: FY 2005–2010

Critical Topic Areas: measurement methods, exposure assessment

Accomplishments and Research Findings (laboratory and field research)

- Collaborated on collection and analysis of diesel particulate samples from mines.
- Collaborated on screening survey at a major producer of carbon nanofibers. Provided analytical data and submitted report to company (screening survey completed).
- Completed walk-through at same facility.
- Completed survey on ultrafine aerosol sources at diesel engine manufacturing plant.
- Completed in-depth survey at the University of Dayton Research Institute carbon nanofiber composite facility. Prepared report to the Institute.
- Provided real-time data to Altair environmental health and safety facilitator for internal review in September 2007. Identified several areas where development of new/improved monitoring approaches is required.

- Completed construction of a novel instrument for fast measurement of nanoparticle size distribution (instrument evaluation under way).

Publications and Abstracts

- Heitbrink WA, Evans DE, Peters TM, Slavin TJ [2007]. Characterization and mapping of very fine particles in an engine machining and assembly facility. J Occup Environ Hyg *4*(5):341–351.
- Baron PA, Estill CF, Beard JK, Hein MJ, Larsen L [2007]. Bacterial endospore inactivation caused by outgassing of vaporous hydrogen peroxide from polymethyl methacrylate (Plexiglas®). Lett Appl Microbiol *45*:485–490.
- Kulkarni P, Deye G, Baron P [2008]. Bipolar diffusion chargin of carbon nanotube aerosols. J Aerosol Sci, doi:10.1016/j.jaerosci.2008.09.008.
- Wang Z, Hopke PK, Ahamadi G, Cheng Y-S, Baron PA [2008]. Fibrous particle deposition in human and nasal passage: the influence of particle length, flow rate, and geometry of nasal airway. J Aerosol Sci *39*(12):1040–1054.
- Baron P, Deye GJ, Martinez AB, Jones EN, Bennett JS [2008]. Size shifts in measurements of droplets with the aerodynamic particle sizer and the aerosizer. Aerosol Sci Technol *42*(3):201–209.
- Olfert J, Kulkarni P, Wang J [2008]. Rapid measurements of aerosol size distributions using a fast integrated mobility spectrometer. J Aerosol Sci [www.ecd.bnl.gov/pubs/BNL-79289-2007-AB.pdf].
- Evans DE, Heitbrink WA, Slavin TJ, Peters TM [2008]. Ultrafine and respirable particles in an automotive grey iron foundry. Ann Occup Hyg *52*(1):9–21.
- Baron PA, Estill CF, Deye GJ, Hein MJ, Beard JK, Larsen LD, Dahlstrom GE [2008]. Development of an aerosol system for uniformly depositing bacillus anthracis spore particles on surfaces. Aerosol Sci Technol *42*(3):159–172.
- Shvedova AA, Kisin E, Murray AR, Johnson V, Gorelik O, Arepalli S, Hubbs AF, Mercer RR, Stone S, Frazer D, Chen T, Deye G, Maynard A, Baron P, Mason R, Kadiiska M, Stadler K, Mouithys-Mickalad A, Castranova V, Kagan VE [2008]. Inhalation of carbon nanotubes induces oxidative stress and cytokine response causing respiratory impairment and pulmonary fibrosis in mice. Toxicologi *102*(1):307.
- Baron PA, Deye GJ, Chen B, Schwegler-Berry DE, Shvedova AA, Castranova V [accepted for publication]. Aerosolization of single-walled carbon nanotubes for an inhalation study. J Inhal Toxicol.
- Shvedova AA, Kisin E, Murray AR, Johnson V, Gorelik O, Arepalli S, Hubbs AF, Mercer RR, Keohavong P, Sussman N, Jin J, Stone S, Chen BT, Deye G, Maynard A, Castranova V, Baron PA, Kagan VE [in press]. Inhalation versus aspiration of single-walled carbon nanotubes in C57BL/6 mice: inflammation, fibrosis, oxidative stress, and mutagenesis. Amer J Physiol.
- Heitbrink WA, Evans DE, Ku B-K, Maynard AD, Slavin TJ, Peters TM [2009]. Relationships between particle number, surface area, and respirable mass concentration in automotive engine manufacturing. J Occup Environ Hyg. *6*(1):19–32.

- Kulkarni P, Deye G, Baron P [in press]. Bipolar diffusion charging characteristics of airborne carbon nanotube aerosols. J Aerosol Sci.

Invited Presentations

- Birch ME, Evans DE, Ku B-K [2007]. Occupational monitoring of carbonaceous nanomaterials. American Association for Aerosol Research annual conference, Reno, Nevada, September 24–28.
- Heitbrink WA, Evans DE, Ku B-K, Maynard AD, Peters TM, Slavin TJ [2007]. The relationship between particle surface area, number, and respirable mass concentration in an automotive foundry and engine machining facility. American Association for Aerosol Research annual conference, Reno, Nevada, September 24–28.
- Kulkarni P, Deye G, Baron P [2007]. Bipolar diffusion charging characteristics of airborne, single-walled carbon nanotubes. American Association for Aerosol Research annual conference, Reno, NV, September 24–28.
- Evans DE [2007]. Working at the Nanoscale: The occupational health and safety perspective. NANOMIST meeting, Birmingham, United Kingdom, February 19.
- Baron PA [2008]. Aerosol science enabling nanomaterial research: experience in environmental and occupational health. Inaugural meeting of the student chapter of American Association for Aerosol Research, University of Cincinnati, Cincinnati, OH, April 29.

Project 25. Measurement of Nanoscale Carbonaceous Aerosols

Principle Investigator M. Eileen Birch, Ph.D.
Project Duration FY 2007–2008

Critical Topic Areas: measurement methods, exposure assessment

Accomplishments and Research Findings (laboratory and field research)

- Completed preliminary investigations at two carbon nanofiber (CNF) facilities when this project began in June 2007. A NIOSH Health Hazard Evaluation (HHE) report on the first field study was published in 2006. A report on the second study was sent to the company in February 2007.
- Completed design and assembly of a mobile aerosol sampling unit (cart) for performing real-time aerosol measurements in workplaces. The unit is equipped with sampling ports and battery power and accommodates a laptop and multiple aerosol instruments.
- Completed in-depth field study (five site visits during FY 2007 and FY2008) at a major CNF manufacturer. Found CNF dispersion within the facility. Identified production byproducts and informed company of them. Also provided recommendations to improve air quality.

- Began work on generation of a carbon black aerosol and its collection on filter sets in July 2008. This work will continue in FY 2009 under another project.
- Participated in the American Ceramics Society Workshop, nanoparticle measurement needs for environmental health and safety in Arlington, Virginia, June 9 and 10, 2008.

Publications and Abstracts

- Methner MM, Birch ME, Evans DE, Ku B-K, Crouch KG, Hoover MD [2007]. Case study: identification and characterization of potential sources of worker exposure to carbon nanofibers during polymer composite laboratory operations. J Occup Environ Hyg *4*(12):D125–D130.
- Noll J, Birch ME [2008]. Effects of sampling artifacts on occupational samples of diesel particulate matter. Environ Sci Technol *42*(14):5223– 5228.
- Kulkarni P, Deye G, Baron P [2008]. Bipolar diffusion charging of carbon nanotube aerosols. J Aerosol Sci, doi:10.1016/j.jaerosci.2008.09.008.
- Wang Z, Hopke PK, Ahamadi G, Cheng Y-S, Baron PA [2008]. Fibrous particle deposition in human and nasal passage: the influence of particle length, flow rate, and geometry of nasal airway. J Aerosol Sci *39*(12):1040–1054.

Invited Presentations

- Birch ME, Evans DE, Ku B-K [2007]. Occupational monitoring of carbonaceous nanomaterials. American Association for Aerosol Research annual conference, Reno, NV, September 24–28.
- Evans DE, Birch ME, Ku B-K, Ruda-Eberenz T [2008]. Occupational monitoring of carbonaceous nanomaterials. American Association for Aerosol Research annual conference, Orlando, FL, October 20–24.
- Ruda-Eberenz T, Birch ME, Evans DE, Ku B-K [2008]. Occupational monitoring of carbonaceous nanomaterials. Central Regional meeting of the American Chemical Society, Columbus, OH, June 10–14.

Project 26. Nanoparticle Reference Materials for Health Protection

Principle Investigator: Aleksandr B. Stefaniak, Ph.D.
Project Duration: FY 2007–2011

Critical Topic Area: measurement methods

Accomplishments and Research Findings

- Developed a project protocol in collaboration with the National Institute of Standards and Technology (NIST) and Oak Ridge National Laboratory to develop reference materials.

- Obtained external peer-review of study protocol.
- Identified researchers at NIST who are leading nanoparticle reference material development initiatives.
- Hosted Vince Hackley (NIST) at NIOSH-Morgantown to discuss NIOSH contributions to the development and qualification of a nanoscale TiO_2 reference material.
- Identified a single-walled carbon nanohorn as an engineered nanoparticle with potential for development into reference material as a control material for toxicology studies.
- Conducted a telephone conference call of NTRC working group on engineered nanoparticle classification schemes.
- Participated in NIST-sponsored workshop on nanomaterial manufacturing, May 20–22, 2008 in Gaithersburg, MD.
    - Participated in breakout session on separation and fractionation of nanomaterials.
    - Reported output of workshop to the National Nanotechnology Initiative.
- Participated in the American Ceramics Society workshop on nanoparticle measurement needs for environmental health and safety, June 9 and 10, 2008 in Arlington, VA.
- Co-sponsored with American National Standards Institute, National Cancer Institute, and NIST a workshop on enabling standards for nanomaterial characterization, October 8 and 9, 2008 in Gaithersburg, Maryland. Acted as codiscussion leader of breakout session—Reference and test materials, media and sample preparation.
- Visited Luna NanoWorks (Danville, Virginia) to discuss workplace measurement issues.

Publications and Abstracts

- Stefaniak AB, Schwegler-Berry D, Hoover MD, Goia DV, Rossner A, Postek MT, Poster DL [2007]. Development of nanoparticle size and surface area reference materials for exposure assessment. International symposium on occupational health implications of nanomaterials, Taipei, Taiwan, August.
- Mark D, Bard D, Wake D, Hoover MD, Stefaniak AB, Methner MM, Evans DE, Geraci CL [2008]. Problems and experience of measuring worker exposure to engineered nanoparticles. International inhalation symposium: benefits and risks of inhaled engineered nanoparticles, Hanover, Germany, June 11–14.

Invited Presentations

- Stefaniak AB [2008]. Methodological challenges of assessing bioavailability of emerging contaminants. University of North Carolina Super- fund Basic Research Program workshop, assessing bioavailability as a determinant of pollutant exposure: building a multidisciplinary paradigm for the 21st century and beyond, Tampa, FL, February 18–20.

- Stefaniak AB [2008]. Standards needs for occupational exposure assessment of nanomaterials. National Institute of Standards and Technology workshop on enabling standards for nanomaterial characterization, Gaithersburg, MD, October 8 and 9.
- Stefaniak AB (presented by Hoover MD) [2008]. Minimal material characterization of nanomaterials. Workshop on ensuring appropriate material characterization in nanotoxicity studies, Washington, DC, October 28.

Project 27. Calm Air Chamber and Wind Tunnel Evaluation of Personal Aerosol Samplers for Nanoparticle Exposure Assessment

Principal Investigator: Terri Pearce, Ph.D.
Project Duration: FY 2008–2010

Critical Topic Area: measurement methods

Accomplishments and Research Findings

- Developed study protocol.
- Obtained external peer-review for protocol.

Invited Presentation

- Pearce T [2008]. Direct-reading nanoaerosol instrument comparison. Abstract and poster for the American Association for Aerosol Research annual conference, Orlando, FL, October 20–24.

Project 28. Ultrafine TiO_2 Surface and Mass Concentration Sampling Method

Principle Investigator: Aleksandr B. Stefaniak, Ph.D.
Project Duration: FY 2007–2011

Critical Topic Area: measurement methods

Accomplishments and Research Findings

- Obtained external peer-review of study protocol.
- Obtained bulk ultrafine TiO_2 powder as a study material.
- Characterized bulk powder physicochemical properties (density, morphology, primary particle size, particle cluster size, total surface area by gas adsorption, purity, and crystallinity).

- Characterized intra- and inter-filter variability for three media: track-etched polycarbonate (0.4 μm pore), track-etched polycarbonate (0.8 μm pore) and Teflon (5 μm pore) including thermal integrity, pressure drop, mass stability, and collection efficiency.
- Developed a protocol for determining background surface area of blank filter media using krypton gas isotherms.
- Characterized intra- and inter-filter surface area variability for track-etched polycarbonate (0.4 μm pore), track-etched polycarbonate (0.8 μm pore) and Teflon (5 μm pore) filters.
- Developed protocol for measuring ultrafine TiO_2 surface area with specificity using bulk powder and perylene diimide dye adsorbed to particle surfaces.

Project 29. Standard Determination of Nanoparticle Size

Principle Investigator: Aleksandr B. Stefaniak, Ph.D.
Project Duration: FY 2009–2010

Critical Topic Area: measurement methods

Accomplishments and Research Findings

- Obtained external peer-review of study protocol.
- Procured three different sizes of spherical gold nanoparticle reference materials from the National Institute of Standards and Technology (NIST).
- Developed a protocol for depositing each size of gold nanoparticle reference material onto functionalized electron microscopy grids.
- Determined particle diameter for each size of gold reference material on an electron microscopy grids.
- Evaluated data to compare measurement results with certified size values. Communicated with NIST statisticians on comparison statistics.
- Cosponsored with American National Standards Institute, National Cancer Institute, and NIST a workshop on enabling standards for nanomaterial characterization, Oct 8 and 9, 2008 in Gaithersburg, Maryland. Acted as co-discussion leader of breakout session— Reference and test materials, media and sample preparation.

Project 30. Development and Evaluation of Nanoaerosol Surface Area Measurement Methods

Principal Investigator: Bon-Ki Ku, Ph.D.
Project Duration: FY 2009

Critical Topic Area: measurement methods

Planned Research

- The objectives of this project are to (1) generate well characterized nanoaerosols with different physical and chemical properties to be used as test aerosols and (2) completely characterize diffusion charging and mobility analysis methods for nanoaerosol particles in the size range up to several microns.
- Conduct a preliminary study on diffusion charger responses to spherical particles in the range from 100 nm up to 600 nm to assess whether the diffusion charger response decreases with increasing mobility diameter.

Project 31. Workplace Monitoring of Carbon Nanofibers/Nanotubes

Principle Investigator: M. Eileen Birch, Ph.D.
Project Duration: FY 2009

Critical Topic Areas: measurement methods, exposure assessment

Accomplishments and Research Findings

- Contacted carbon nanotube manufacturer to discuss production processes and tentative date for field survey in FY 2009.

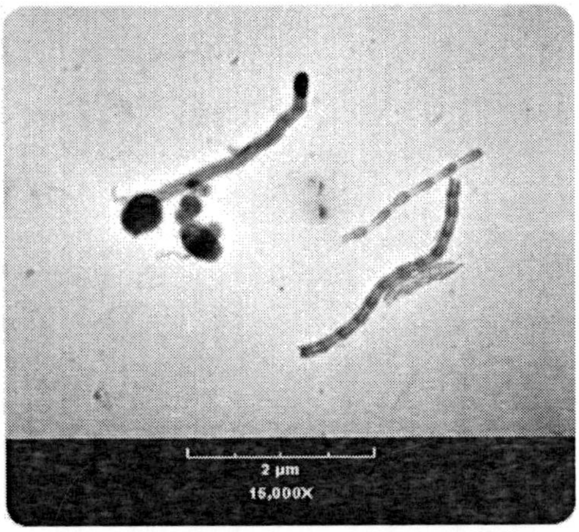

Electron micrograph of carbon nanofibers at 15,000x

- Generation methods for carbon black aerosol are being investigated. The objective is collection of a matched filter set for interlaboratory comparisons.
- Initiated collaboration with NIST on an analytical reference material for carbon nanotubes.

Project 32. Ultrafine Aerosols from Diesel-Powered Equipment

Principal Investigator: Aleksandar Bugarski, Ph.D.
Project Duration: FY 2004–2008

Critical Topic Area: measurement methods

Accomplishments and Research Findings

- Evaluated the effects of a fuel-borne, catalyst-regenerated, sintered metal, diesel-particulate-filter (DPF) system from Mann-Hummel on size distribution and concentration of nano and ultrafine aerosols in mine air. This study conducted at NIOSH Diesel Laboratory at Lake Lynn Laboratory was used to characterize physical and chemical properties of aerosols emitted from an engine equipped with sintered metal DPF system and to assess the effects of DPF regeneration strategy (fuel- borne catalyst) on concentrations of aerosols and gases in mine air.
- Evaluated the potential of biodiesel fuels to reduce concentrations of aerosols in mine air. This study was conducted at the NIOSH Diesel Laboratory at Lake Lynn Laboratory. Compared neat and blended biodiesel fuels with ultra low sulfur petroleum diesel relating to their effects on the concentrations and size distributions of diesel aerosols in underground mine conditions. The physical and chemical properties and toxicity of diesel and biodiesel aerosols in mine air were studied.
- Evaluated the effects of an electrically regenerated, sintered metal DPF system from Rypos on size distribution and concentration of nano- and ultrafine aerosols in mine air. This study, conducted at NIOSH Diesel Laboratory at Lake Lynn Laboratory, was used to characterize physical and chemical properties of aerosols emitted from an engine equipped with sintered metal DPF system and to assess the effects of DPF-regeneration strategy (electrical current) on concentrations of aerosols and gases in mine air.
- Developed novel methodology for collection of diesel particulates for genotoxicity analysis. The samples are collected directly on lung surfactant using biosamplers and versatile aerosol concentration enrichment system developed in cooperation with University of Southern California.
- Developed the novel methodology for preparing and analyzing samples collected using aforementioned sampling system.
- Using the salmonella reversion assays, assessed genotoxic potential of the solvent extracted diesel aerosols samples collected during DPF and biodiesel studies.
- Evaluated suitability of various aerosol instrumentation, such as fast mobility particle sizer, scanning mobility particle sizer, electrical low pressure impactor, nanoparticle surface area monitor, PAS2000, and filter-sampling methodologies for measurement and sampling of diesel aerosols.
- Completed the new engine/dynamometer system with a Mercedes- Benz engine and 400 kW dynamometer. It is awaiting deployment to NIOSH Diesel Laboratory at Lake Lynn Laboratory.

- The Lake Lynn Laboratory mine has been under reconstruction since July 2008. For safety reasons, the experimental work is postponed probably until spring 2009.
- Disseminated findings in peer-review papers and at 13th and 14th Mining Diesel Emissions Council conference, 13th Diesel Engine- Efficiency and Emissions Research (DEER) conference, 12th United States/North American mine ventilation symposium.

Invited Presentations

- Bugarski AD, Schnakenberg GH Jr., Hummer JA, Cauda E, Janisko SJ, Patts LD [2007]. Examination of diesel after-treatment systems at NIOSH Lake Lynn Laboratory. Annual mining diesel emissions conference, Richmond Hill/Toronto, Ontario, Canada, October 1–5.
- Shi XC, Keane M, Ong T, Harrison J, Gautam M, Li S-Q, Bugarski A, Wallace W [2007]. Preserving diesel exhaust ultrafine (nano-) particulate structure in genotoxicity studies to support engineering development of emissions controls. Diesel engine-efficiency and emissions research (DEER) conference, Detroit, MI, August 13–16.
- Bugarski A, Schnakenberg GH Jr., Mischler S, Cauda E [2007]. Characterization of the physical and chemical properties of diesel aerosols in an underground mine diesel laboratory. ETH conference on combustion generated nanoparticles, Zurich, Switzerland, August 13–15.
- Gautam M, Wilt G, Carder D, Bugarski A [2007]. Evaluation of diesel exhaust particulate matter in a ventilated mine tunnel. ETH conference on combustion generated nanoparticles, Zurich, Switzerland, August 13–15.
- Bugarski AD, Cauda E, Janisko S, Patts LD, Hummer JA, Mischler SE [2008]. Effects of biodiesel on aerosols in underground mine. Annual mining diesel emissions conference, Richmond Hill/Toronto, Ontario, Canada, October 5–10.
- Bugarski AD, Cauda E, Janisko S, Patts LD, Hummer JA, Mischler SE [2008]. Evaluation of an electrically regenerated sintered metal diesel particulate filter system in underground mine laboratory. Annual mining diesel emissions conference, Richmond Hill/Toronto, Ontario, Canada, October 5–10.
- Shi X-C, Keane MJ, Ong T-M, Bugarski AD, Gautam M, Wallace WE [2008]. Diesel exhaust particulate material expression of in vitro genotoxic activities when dispersed into a phospholipid component of lung surfactant. Inhaled Particles X conference, Sheffield, United Kingdom, September 23–25.
- Bugarski AD, Schnakenberg GH Jr., Cauda E [2008]. Effects of sintered metal diesel particulate filter system on diesel aerosols and nitric oxides in mine air. United States/North American mine ventilation symposium, Reno, NV, June 9–11.
- Bugarski A [2008]. Diesel particulate filtration technology in underground mines. California Air Resource Board (CARB)/Southern California Air Quality Management District, course on ultrafine particles and retrofit technologies for diesel engines, Diamond Bar, CA, November 12–14.

Project 33. Titanium Dioxide and Other Metal Oxides Exposure Assessment Study

Principle Investigator: Brian Curwin, Ph.D.
Project Duration: FY 2006–2009

Critical Topic Areas: measurement methods, exposure assessment

Accomplishments and Research Findings

- Wrote protocol for titanium dioxide study, which underwent peer review.
- Purchased critical exposure assessment equipment, including a TSI 3007 condensation particle counter, an Ecochem DC2000 CE portable diffusion charger, and a MOUDI impactor.
- Awarded a contract for support in company recruitment. Recruitment was completed in June 2008.
- Contacted 189 companies for recruitment; 10 initially agreed to participate and surveys were conducted at 7 companies. Three more companies waiting to be scheduled.
- Sent survey reports to three companies.

Project 34. Field Research Team

Principle Investigators: Mark Methner, Ph.D., and Charles Geraci, Ph.D.
Project Duration: FY 2006–2012

Critical Topic Area: exposure assessment

Accomplishments and Research Findings (field surveys)

- Performed sampling at a multi-walled carbon nanotubes manufacturer. Report issued in 2007.
- Completed two site visits at a metal oxide manufacturer to evaluate engineering controls (2007, 2008). Reports issued in 2008. Engineering case study published in the Journal of Occupational and Environmental Health in 2008.
- Conducted an observational walkthrough at a research and development laboratory that handles and compounds nanoscale polyhedral oligomeric silsesquioxanes (POSS). Subsequent air sampling was performed by facility. Future collaborations on additional air sampling and possible engineering controls are expected.
- Performed sampling at a manufacturer of nanoenhanced, silica-iron, absorbent media. Evaluated a spray deposition of nanomaterial onto cellulose substrate inside ventilated chamber. Report issued in 2008.
- Completed two trips to evaluate chamber ventilation role in controlling releases during cleanout operations at a pilot-scale research and development facility that

creates nanoscale alumina via radiofrequency plasma torch reaction. Issued two reports (2007, 2008). Future collaboration to evaluate alternative engineering controls expected.

- Performed sampling at an environmental toxicology laboratory. Facility conducts toxicology studies on invertebrates. Evaluated handling procedures during MWCNT- and fullerene-solution preparation. Report issued in 2008.
- Performed sampling at a manufacturer that compounds/blends MWCNTs and boron carbide. Company ships product to manufacturer that uses material to create body armor. Report issued in 2009. Expect future collaborations to include engineering control evaluation and potential exposure to end-user of material.
- Performed sampling at a pilot-scale, research-and-development laboratory that handles/processes MWCNTs for use in composite material production. Facility installed HEPA-filtered downdraft ventilated enclosure for use during handling operations. Air sampling conducted with and without ventilation system operating. Report issued in 2009.
- Performed sampling at a research and development laboratory that handles a variety of engineered nanomaterials (MWCNTs, metal oxides). Report issued in 2009.
- Performed sampling at a pilot-scale, research-and-development facility that synthesizes and handles MWCNTs for inclusion in composite materials. Report issued in 2009.

Cleaning of metal oxide reactor. Note local exhaust ventilation is in use. *Image courtesy of Mark Methner, NIOSH.*

- Performed sampling at a Department of Defense Applied Toxicology laboratory that handles engineered nanomaterials (elemental metals and oxides) for use in animal toxicological studies. Report issued in 2009.
- Performed sampling at a research laboratory producing carbon nano- tube infused fiber composites. Report issued in 2009.

Publications and Abstracts

- Methner MM, Birch ME, Evans DE, Ku B-K, Crouch K, Hoover MD, Mazzukelli LF (Ed) [2007]. Case study: identification and characterization of potential sources of worker exposure to carbon nanofibers during polymer composite laboratory operations. J Occup Environ Hyg *4*(12):D125–D130.
- Methner MM and Old L (Ed.) [2008]. Engineering case study: effectiveness of local exhaust ventilation in controlling engineered nanomaterial emissions during reactor cleanout operations. J Occup Environ Hyg *5*(6):D63–69.

Invited Presentations

- Hodson LL [2007]. Nanoparticles-NIOSH Nanoparticles Research Program. The Insurance Loss Control Association, Oct 31.
- Hodson, LL [2008]. Workplace exposure assessment and risk management practices for nanomaterials. The Ohio Safety Congress and Expo, Columbus, OH, April 3.
- Methner MM, Hodson LL, Geraci CL [2008]. One day workshop: Nanoparticles in the Workplace: Demonstration of the instruments used when using the nanoparticle emission assessment technique (NEAT). American Industrial Hygiene Association, Central Ohio Section, Columbus, OH, April 22.
- Hodson LL [2008]. Nanoparticles—NIOSH Nanoparticle Research Program. The Edison Electric Institute spring occupational health and safety conference, Louisville, KY, April 29.
- Methner MM [2008]. Nanotechnology: ask the expert panel discussion: results pertaining to the field use of the nanoparticle emission assessment technique (NEAT). American Industrial Hygiene Association conference and exposition, Minneapolis, MN, May 31–June 5.
- Methner MM, Hodson LL [2008]. Professional Development Course: Nanoparticles in the Workplace: Demonstration of the instruments used when using the nanoparticle emission assessment technique (NEAT). American Industrial Hygiene Association conference and exposition, Minneapolis, MN, May 31–June 5.
- Mark D, Bard D, Wake D, Hoover MD, Stefaniak AB, Methner MM, Evans DE, Geraci CL [2008]. Problems and experiences of measuring worker exposure to engineered nanoparticles. International Inhalation Symposium, Hanover, Germany, June 11–14.
- Hodson LL [2008]. The NIOSH field research team—a brief summary and update of activities. U.S. Air Force (USAF) workshop on biological interaction of engineered nanoparticles, Wright Patterson AFB, Dayton, OH, June 24.
- Hodson LL [2008]. NIOSH Nanotechnology Research Program. American Industrial Hygiene Association, Indiana section meeting, Indianapolis, IN, July 9.
- Methner MM, Hodson LL, Geraci CL [2008]. Professional Development Course: Nanoparticles in the Workplace: Demonstration of the instruments used when using the nanoparticle emission assessment technique (NEAT). American Industrial Hygiene Association, Ohio Valley Section, West Chester, OH, December 16.

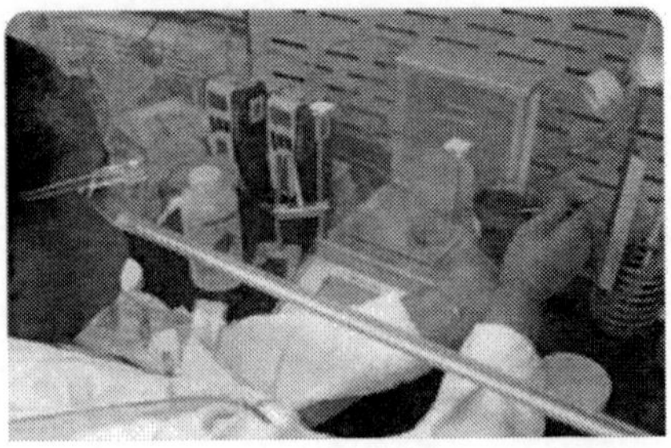

Weighing CNF in an exhausted enclosure. *Image courtesy of Mark Methner, NIOSH.*

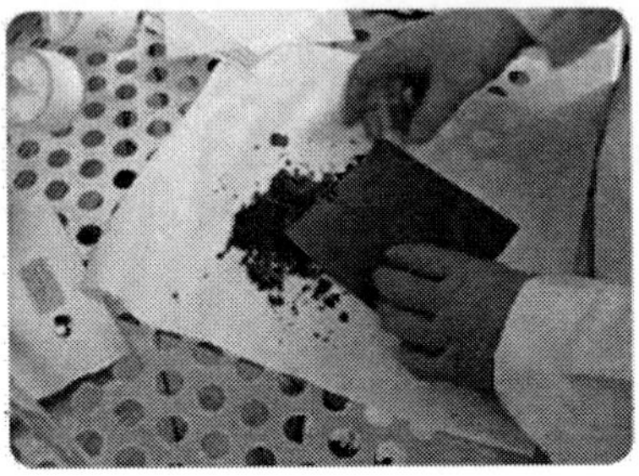

Crumbling carbon nano- fiber paper. *Image courtesy of Mark Methner, NIOSH.*

Project 35. Assessing the Feasibility of Industrywide Exposure and Epidemiology Studies of Workers Exposed to Engineered Nanomaterials

Principle Investigator: Mary Schubauer-Berigan, Ph.D.
Project Duration: FY 2008–2009

Critical Topic Area: epidemiology and surveillance

Accomplishments and Research Findings

- Developed and funded a statement of work to collect and compile information on the size, characteristics, and future trends of the U.S. workforce involved in the manufacture of engineered carbonaceous nanomaterials.
- Attended the 2008 Micro Nano Breakthrough conference in Vancouver, Washington, to learn about the latest applications of engineered carbonaceous nanomaterials as well as possible timeframes and barriers for transition from startup to full-scale production for the most promising applications.
- Reviewed and compiled information on companies identified as engineered carbonaceous nanomaterials manufacturers in the Lux Report (fourth and fifth editions), supplemented with information collected from Dun and Bradstreet.
- Contacted and conducted interviews with health and safety managers at each identified company to confirm the information collected on workforce size, location,

characteristics of the nanomaterials produced, and use of engineering controls and personal protective equipment.

- Developed a report on the feasibility of conducting industrywide exposure surveys and epidemiological studies among this workforce.

Project 36. Nanoparticles—Dosimetry and Risk Assessment

Principle Investigator: Eileen D. Kuempel, Ph.D.
Project Duration: FY 2005–2012

Critical Topic Area: risk assessment

Accomplishments and Research Findings

- Developed and published risk assessment methods for inhaled poorly soluble nanoparticles, including quantitative dose-response analysis in rodents, dosimetry modeling, and extrapolation of risk estimates to humans.
- Completed nanoparticle dosimetry model in rats. Dosimetry model describes clearance, retention, and translocation from lungs to systemic circulation, and early biological responses. Model calibration used dose-response data from NIOSH animal studies (IOM/NIOSH research collaboration).
- Completed olfactory deposition model of spherical nanoparticles in rats. Revised multipath particle deposition (MPPD) model to include site-specific deposition efficiency in nasal-pharyngeal region (Hamner Institute/NIOSH research collaboration).
- Initiated research with Hamner Institute to develop human olfactory deposition model of spherical nanoparticles. Will use estimates of olfactory deposition proportion in estimating dose and assessing risk of nanoaerosols.
- Developed modeling methods to predict human respiratory tract deposition model for carbon nanofibers/nanotubes structures based on fiber aerosol theory. Additional methods are being evaluated to account for diffusion diameter and to model deposition of heterogeneous nano- aerosol structures. (Hamner Institute/NIOSH research collaboration).

Publications and Abstracts

- Kuempel ED [2007]. Estimating nanoparticle dose in humans: issues and challenges. Chapter 10. In: Monteiro-Riviere NA, Tran CL, eds. Nanotoxicology: characterization, dosing, and health effects. New York: Informa Healthcare USA Inc., pp. 141–152.
- Kuempel ED, Geraci CL, Schulte PA [2007]. Risk assessment approaches and research needs for nanoparticles: An examination of data and information from current studies. In: Simeonova PP, Opopol N, Luster MI, eds. Nanotechnology:

toxicological issues and environmental safety, NATO security through science series book. New York: Springer-Verlag, pp. 119–145.

- Dankovic D, Kuempel E, Wheeler M [2007]. An approach to risk assessment for TiO_2. Inhal Toxicol *19*(Suppl 1):205–212.
- Tran CL, Kuempel ED [2007]. Biologically based lung dosimetry and exposure-dose-response models for poorly soluble, inhaled particles. In: Donaldson K, Borm P, eds. Particle Toxicology. Boca Raton, Florida: CRC Press, Taylor and Francis Group, Chapter 20.
- Kuempel ED, Tran CL, Castranova V, Bailer AJ [2007]. Response to letter to the editor from Dr. Peter Morfeld [re: Lung dosimetry and risk assessment of nanoparticles: evaluating and extending current models in rats and humans. Inhal Toxicol *18*(10):717–24]. Inhal Toxicol *19*(2): 197–198.
- Garcia GJM, Nazridoust K, Kimbell JS [2008]. Nanoparticle deposition in the rat nasal cavity: prediction of dose to the olfactory epithelium. Toxicologist *102*(1):204.
- MacCalman L, Tran CL, Kuempel E [in press]. Development of a biomathematical model in rats to describe particle size-specific clearance and translocation of inhaled nano particles. J Physics.
- Kuempel ED, Smith RJ, Dankovic DA, Stayner LT [in press]. Rat- and human-based risk estimates of lung cancer from occupational exposure to poorly soluble particles: a quantitative evaluation. J Physics.

Invited Presentations

- Kuempel ED [2007]. Review of hazard information on nanoparticles: as relates to occupational health surveillance. Workshop on occupational health surveillance and nanotechnology workers. Washington, DC, April 17.
- Geraci CL, Schulte PA, Zumwalde R, Kuempel E, Hoover M [2007]. Occupational risk management for nanomaterials along the product life cycle. International symposium on nanotechnology, occupational, and environmental health. Taipei, Taiwan, August 29–September 1.
- Schulte PA, Murashov VV, Geraci CL, Zumwalde RD, Hoover MD, Castranova V, Kuempel ED [2007]. Current occupational safety and health issues of nanotechnology in the USA. European NanOSH conference—Nanotechnologies: a critical area in occupational safety and health, Helsinki, Finland, December 3–5.
- MacCalman L, Tran CL, Kuempel E [2008]. Development of a biomathematical model in rats to describe the retention, clearance, and translocation of nanoparticles from the lungs. Inhaled Particles X conference, Sheffield, United Kingdom, September 23–25.
- Kuempel ED, Smith RJ, Dankovic DA, Stayner LT [2008]. Rat- and human-based risk estimates of lung cancer from occupational exposure to poorly soluble particles: a quantitative evaluation. Inhaled Particles X conference, Sheffield, United Kingdom, September 23–25.
- MacCalman L, Tran L, Kuempel E [2008]. Developing and testing a biomathematical model to describe particle size-specific clearance and translocation

of nanoparticles in rats. Nanotoxicology—Second international nanotoxicology conference, Zurich, Switzerland, September 7–10.

- Schulte PA, Kuempel E, Castranova V, Trout D [2008]. Assessing the toxicological evidence base for medical surveillance of workers potentially exposed to engineered nanoparticles. Second international nano- toxicology conference, Zurich, Switzerland, September 7–10.
- Kuempel ED [2008]. Should engineered nanoparticles be considered "new chemicals"? NTRC annual meeting, Morgantown, West Virginia, January 23.

Project 37. Penetration of Nanoparticles through Respirators

Principle Investigator: Samy Rengasamy, Ph.D.
Project Duration: FY 2005–2009

Critical Topic Area: engineering controls and personal protective equipment

Accomplishments and Research Findings

- Conducted studies to measure nanoparticle penetration through the following:
  1. NIOSH-approved N95 and P100 and European-certified CE marked filtering facepiece respirators
  2. Dust masks that are available in home improvement/hardware stores
  3. FDA-approved surgical masks

Publications and Abstracts

- Rengasamy S, Verfobsky R, King W, Shaffer R [2007]. Nanoparticle penetration through NIOSH-approved N95 filtering facepiece respirators. JISRP *24*:49–59.
- Rengasamy S, Eimer B, Shaffer R [2008]. Nanoparticle filtration performance of commercially available dust masks. JISRP *25*:27–41.
- Rengasamy S, King W, Eimer B, Shaffer R [2008]. Filtration performance of NIOSH-approved N95 and P100 filtering facepiece respirators against 4–30 nanometer-size nanoparticles. J Occup Environ Hyg *5*:556–564.
- Rengasamy S, Eimer B, Shaffer R [in press]. Comparison of nanoparticle filtration performance of NIOSH-approved and CE-marked filtering facepiece respirators. Annals of Occ Hyg.
- Shaffer R, Rengasamy S [in press]. Respiratory protection against nanoparticles: A review. J Nanopart Res. Special issue on "Exposure assessment and exposure mitigation for nanomaterials in the workplace".
- Fisher E, Rengasamy S, Viscusi D, Vo E, Shaffer R [in press]. Development of a test system of apply virus containing particles to filtering facepiece respirators for the evaluation of decontamination procedures. App Environ Microbiol.

Invited Presentations

- Shaffer R [2007]. Air purifying respirators for the nanotechnology industry. National Response Team-Worker Safety and Health Conference, Washington, DC, August 7.
- Shaffer R [2007]. An overview of NIOSH nanotechnology research and an update on the efficacy of personal protective equipment for reducing worker exposure to nanoparticles. Commercialization of nanomaterials meeting, Pittsburgh, PA, November 11–13.
- Rengasamy S, Eimer B, Shaffer R [2008]. Filtration performance of NIOSH-approved N95 and P100 filtering facepiece respirators against nanoparticles. American Industrial Hygiene Association conference, Minneapolis, MN, May 31–June 5.
- Rengasamy S, Eimer B, Shaffer R [2008]. Comparison of nanoparticle filtration performance of NIOSH-approved and European-certified filtering facepiece respirators. ISRP international conference, Dublin, Ireland, September 14–18.
- Rengasamy S, Eimer B, Shaffer R [2008]. Comparison of nanoparticle filtration performance of NIOSH-approved and European-certified filtering facepiece respirators. American Association for Aerosol Research annual conference, Orlando, FL, October 20–24.
- Rengasamy S, Shaffer R [2007]. Respiratory protection against nanoparticles. Chesapeake Area Biological Association scientific symposium, Columbia, MD, June 14.
- Shaffer R [2008]. Effectiveness of personal protective equipment and engineering controls. Safe handling of engineered nanoscale materials symposium, Argonne, IL, July 7 and 8.
- Shaffer R [2008]. Control and filtration of nanomaterials. Occupational health aspects of nanotechnology symposium. Northern California Section of the American Industrial Hygiene Association, Menlo Park, CA, May 14.
- Rengasamy S [2008]. Respiratory protection against biological and inert particles. International Society for Respiratory Protection meeting on respiratory protection issues in emergency response and healthcare, Montreal, Canada, October 6.
- Rengasamy S [2008]. Respiratory protection against nanoparticles. Institut de recherche en sante et en securite du travail, colloquium on respiratory protection: research priorities, Montreal, Canada, October 7.

Project 38. Nanoparticle Penetration through Protective Clothing

Principle Investigator: Pengfei Gao, Ph.D.
Project Duration: FY 2009–2009

Critical Topic Area: engineering controls and personal protective clothing

Planned Research

- Development of a model based on single-fiber theory to predict nanoparticle penetration through clothing materials.
- Development of a bench-scale particulate penetration test method based on a wind-driven approach.
- Measurement of particle penetration levels through various fabrics used in protective clothing ensembles.

Project 39. Development of Personal Protective Equipment Ensemble Test Methods

Principle Investigator: Pengfei Gao, Ph.D.
Project Duration: FY 2005–2009

Critical Topic Area: engineering controls and personal protective equipment

Accomplishments and Research Findings

- Developed a multidomain, magnetic passive aerosol sampler for measuring particulate penetration through protective ensembles. Submitted an employee invention report to CDC Technology Transfer Office: Ref. I–007–08 (Gao P [2007]).

Publications and Abstracts

- Gao P, King WP, Shaffer R [2007]. Review of chamber design requirements for testing of personal protective clothing ensembles. J Occup Environ Hyg *4*(8):562–571.
- Gao P, Jeffrey LB, Shaffer R [in press]. Considerations for selection of PPE to protect against nanoparticle exposure. In: Anna D, ed. Chemical protective clothing, 3rd edition. Fairfax, VA: AIHA.

Invited Presentations

- Wang ZM, Gao P [2007]. A study on magnetic passive aerosol sampler for measuring aerosol particle penetration through protective ensembles. American Association for Aerosol Research annual conference, Reno, NV, September 24–28.
- Shaffer R [2007]. An overview of NIOSH nanotechnology research and an update on the efficacy of personal protective equipment for reducing worker exposure to nanoparticles. Commercialization of nanomaterials meeting, Pittsburgh, PA, November 12.
- Shaffer R [2008]. Control and filtration of nanomaterials. Occupational health aspects of nanotechnology symposium, Northern California Section of the American Industrial Hygiene Association, Menlo Park, CA, May 14.

- Shaffer R [2008]. Effectiveness of PPE and engineering controls. Safe handling of engineered nanoscale materials symposium, Argonne, IL, July 7 and 8.
- Jaques PA, Gao P [2008]. Effect of wind velocity on particle collection using a multidomain magnetic passive aerosol sampler. American Association for Aerosol Research annual conference, Orlando, FL, October 20–24.

Project 40. Web-based Nanoparticle Information Library Implementation

Principal Investigator: Arthur Miller, Ph.D.
Project Duration: FY 2004–2008

Critical Topic Area: communication and information

Accomplishments and Research Findings

- Collaborated with Oregon State University and a Health Effects Institute (HEI) consortium toward the development of an informational hub for nanotechnology and health effects.
- Developed Web links for the NIOSH Nanoparticle Information Library to online resources at the Wilson International Center for International Scholars, the International Council on Nanotechnology bibliography on nanotechnology safety and health at Rice University and the National Science Foundation's Nanomanufacturing Network at the University of Massachusetts at Lowell.
- Continued input for redesigning the nanotechnology topic page on the NIOSH Web site.
- Developed a prototype, portable, hand-held electrostatic precipitator that would enhance development of new information for the Nanoparticle Information Library by enabling nanoparticle sampling and subsequent electron microscopy analysis.
- Developed protocols for analyzing workplace nanoaerosol samples using transmission electron microscopy/energy dispersive spectroscopy.
- Developed a protocol for nanoaerosol characterization fieldwork, including quasi-real time spatial mapping of nanoparticle concentrations in the workplace.
- Updated Nanoparticle Information Library software to meet new CDC security requirements.
- Incorporated a mailing interface and a newsletter interface in the design of the Nanoparticle Information Library.
- Developed online newsletters and disseminated them to persons on the Nanoparticle Information Library mailing list.
- Developed course curriculum entitled Nanotechnology Science and Engineering for the mechanical engineering department at Gonzaga University. The course covers the basic chemical, physical, and engineering principles governing nanomaterials, highlighting their differences from classic materials, as well as the science required to use them in nanotechnologies for real-world problem solving.

- Led a team of students in the design, building, and testing of a rapid mobility particle sizer, which has application for the Nanoparticle Information Library and nanotechnology science in general and won first place in the 2008 American Society of Mechanical Engineers (ASME) national design contest.

Publications and Abstracts

- Miller AL, Stipe C, Habjan MC, Ahlstrand G [2007]. Role of lubrication oil in particulate emissions from a hydrogen-powered internal combustion engine. Environ Sci Technol *41* :6828–6835.
- Miller AL, Hoover M D, Mitchell DM, Stapleton BP [2007]. The nanoparticle information library (NIL): a prototype for linking and sharing emerging data. J Occup Environ Hyg *4*(12):D131–D134.
- Miller A, Ahlstrand G, Kittelson DB, Zachariah MR [2007]. The fate of metal (Fe) during diesel combustion: morphology, chemistry, and formation pathways of nanoparticles. Combust Flame *149*:129–143.
- Miller AL, Habjan MC, Park K [2007]. Real-time estimation of elemental carbon emitted from a diesel engine. Environ Sci Technol *41*:5783– 5788.
- Ng A, Miller A, Ma H, Kittelson DB [2007]. Comparing measurements of carbon in diesel exhaust aerosols using the aethalometer, NIOSH method 5040 and SMPS. SAE tech paper series 2007–01–0334.
- Miller AL, Habjan M, Beers-Green A, Ahlstrand G [2008]. Microscopic analysis of airborne particles and fibers, a chapter for the new American Conference of Governmental Industrial Hygienists handbook, Air sampling technologies: principles and applications [www.acgih.org/ store/ProductDetail.cfm?id=2015].
- Park K, Kim J-S, Miller AL [2008]. A study on effects of size and structure on hygroscopicity of nanoparticles using a tandem differential mobility analyzer and transmission electron microscopy. J Nanopart Res [http://dx.doi.org/10.1007/s 11051-008-9462-4].
- Choo S-W, Lee D, Lee K-S, Miller AL, Park K, Zachariah M, Zhou L [2008]. Understanding nanoscale phenomena using single particle mass spectrometry and improvement of its performance—a review. Chem Physics Res J *2*(1/2):17.

Invited Presentations

- Miller A [2007]. Occupational safety and health challenges of nano- technology: What we know and what we don't know. Oregon governor's occupational health and safety conference, Portland, OR, March 13.
- Miller A [2007]. Characterization of hazardous pollutants in the workplace. IERC workshop for sound management of hazardous chemicals and sustainable energy, sponsored by the Gwangju Institute of Science and Technology, Gwangju, Korea, October 30–November 3.
- Miller A [2008]. Spatial mapping of industrial hygiene data in the workplace. American Industrial Hygiene Association conference and exhibition, Minneapolis, MN, May 31–June 5.

- Miller A, Geraci CL, Maynard A [2008]. NIOSH's role in improving health and safety in nanotechnology workplaces. Special Libraries Association annual meeting, Seattle, WA, June 15–17.
- Miller A [2008]. NIOSH's role in improving health and safety in nano- technology workplaces: knowledge management of nanotechnology information. Micro-nano breakthrough conference, Portland, OR, September 8–10.

Project 41. Nanoparticles in the Workplace

Principle Investigator: Mark D. Hoover, Ph.D.
Project Duration: FY 2004–2009

Critical Topic Areas: exposure assessment, communication and information

Accomplishments and Research Findings

- Summarized and published issues and approaches for exposure assessment of nanoparticles in the workplace as a chapter in a book on nanotoxicology that involved a spectrum of international contributors. The exposure assessment chapter included a comprehensive formulaic method for assessing and managing determinants of workplace exposure such as material at risk, damage ratio, airborne release fraction, respirable fraction, and control factors.
- Contributed to the development of an American Industrial Hygiene Association (AIHA) document on guidance for conducting control banding analyses. Provided leadership in activities of the AIHA control banding working group and the AIHA nanotechnology working group to develop potential nanotechnology applications of control banding approaches from pharmaceutical and other historical experiences.
- Developed a new ultrafine beryllium oxide standard reference material (SRM) through the collaborations with the National Institute of Standards and Technology (NIST), and U.S. Department of Energy (DOE). SRM 1877 was formally approved and issued for distribution and use in August 2008. This is one of the first reference materials issued in the ultrafine particle size range and will improve calibration and development of sampling and analytical methods and toxicity testing. Round- robin evaluations of analytical methods for beryllium detection in air filtration samples spiked with SRM 1877 are underway under the auspices of the Beryllium Health and Safety Committee. Results of the evaluatons are expected to have relevance for digestion and detection effeciencies of other ultrafine metal oxide particles.
- Developed partnerships with nanotechnology industries, academia, government agencies, labor, and voluntary consensus standard committees to conduct joint research and develop guidelines on working safely with nanomaterials. This included involvement in activities of the American National Standards Institute, ASTM International, the International Standards Organization, and the International Electrotechnical Commission. This included collaboration on development of a new Web site, wiki, and other informatics and knowledge-management capabilities for

sharing information, including through the NIOSH Nanoparticle Information Library and through a nanotechnology community initiative on minimum information for nanomaterial characterization.

Publications and Abstracts

- Hoover MD, Stefaniak AB, Day GA, Geraci CL [2007]. Exposure assessment considerations for nanoparticles in the workplace. In: Monteiro-Riviere NA, Tran CL, eds. Nanotoxicology: characterization, dosing, and health effects. New York: Informa Healthcare USA Inc., pp. 71–83.
- Miller AL, Hoover MD, Mitchell DM, Stapleton BP [2007]. The Nanoparticle Information Library (NIL): a prototype for linking and sharing emerging data. J Occup Environ Hyg *4*(12):D131–D134.
- American Industrial Hygiene Association [2007]. Guidance for conducting control banding analyses. Fairfax, VA: American Industrial Hygiene Association, Document 9–2007.
- Hoover MD, Poster D [2008]. Environmental, health, and safety crosscut issues for nanomanufacturing. In: Instrumentation, metrology, and standards for nanomanufacturing workshop final report. Postek MT, Lyons KW, Ouimette MS, Holdridge GM eds. National Science and Technology Council Interagency Working Group on Manufacturing Research and Development, Washington, DC

Invited Presentations

- Hoover MD [2007]. NIOSH research on applications, safety hazards, and environmental issues for nanotechnology. CSHEMA-PRIZIM regional Environmental Health and Safety seminar, University of Central Florida, Orlando, FL, April 17.
- Hoover MD [2007]. NIOSH experience in assessing practical measurement and control practices for nanoparticles. Midwest Nanotechnology Safety workshop, Madison, WI, May 21.
- Hoover MD [2007]. Case studies in assessing practical measurement and control practices for nanoparticles. American Industrial Hygiene Association aerosol technology committee symposium on nanotechnology: wonders and worries, Philadelphia, PA, June 2.
- Hoover MD [2007]. Nanotechnology health and safety issues. U.S. Department of Energy industrial hygiene and occupational safety special interest group, Philadelphia, PA, June 8.
- Hoover MD [2007]. Nanoparticle Information Library. National Nanomanufacturing Network workshop on nanoinformatics strategies, Arlington, VA, June 12.
- Hoover MD [2007]. Reference material updates and insights from the NIOSH nanotechnology and exposure assessment research programs. NIOSH-NIST information exchange meeting, National Institute of Standards and Technology, Gaithersburg, MD, June 14.

- Hoover MD [2007]. Overview of NIOSH research on environmental health and safety issues for nanotechnology session on health and safety issues associated with nanomaterials. Air and Waste Management Association annual meeting, Pittsburgh, PA, June 27.
- Hoover MD [2007]. Environmental Health and Safety overview— exposure assessment and control for nanoparticles in the workplace: insights from the NIOSH Nanotechnology Research Program. NanoScience+Engineering SPIE Optics+Photonics conference, San Diego, CA, August 30.
- Hoover MD [2007]. Materials necessary for health and occupational exposure studies, workshop on standards for environmental health and safety research needs for engineered nanoscale materials. National Institute of Standards and Technology, Gaithersburg, MD, September 12.
- Hoover MD [2008]. NanoMetrology research at NIOSH, National Institute of Standards and Technology—NIOSH scientific information exchange, National Institute of Standards and Technology, Gaithersburg, MD, January 28.
- Hoover MD [2008]. Insights from the NIOSH Nanotechnology Research Program, panel discussion on accelerators and nanoparticles, Health Physics Society midyear meeting on radiation-generating devices, Oakland, CA, January 29.
- Hoover MD [2008]. NIOSH nanotechnology initiative: remarks on exposure assessment and control and informatics opportunities. NIOSH education and research centers industrial hygiene program directors' meeting, Galveston, TX, February 10.
- Hoover MD [2008]. The collection and characterization of nanomaterials for exposure assessment. American Industrial Hygiene Association—Northern California section technical symposium on the occupational health aspects of nanomaterials, Menlo Park, CA, May 14.
- Hoover MD [2008]. Exposure assessment data needs for emerging issues: a case study for nanotechnology risk assessment and risk management. American Industrial Hygiene Association conference and exposition, Minneapolis, MN, May 31–June 5.
- Hoover MD [2008]. Nanoparticle issues for the health physicist: insights from the NIOSH Nanotechnology Research Program. Annual meeting of the Health Physics Society, Pittsburgh, PA, July 16.
- Hoover MD [2008]. Update on the NIOSH direct-reading methods initiative. Annual meeting of the Health Physics Society, Pittsburgh, PA, July 16.
- Hoover, MD [2008]. Demonstration of the NIOSH prototype Web- based Nanoparticle Information Library. Workshop on international collaboration for nanoEHS informatics, National Institute of Standards and Technology, Gaithersburg, MD, October 10.
- Hoover MD [2008]. Nanotechnology safety issues for radioactive materials: insights from the NIOSH Nanotechnology Research Program, regional inspector counterpart seminar. U.S. Nuclear Regulatory Commission, Region 1, King of Prussia, PA, December 10.

Project 42. New Sensor Technology Development for ESLI

Principle Investigator: Jay Snyder
Project Duration: FY 2002–2010

Critical Topic Area: applications

Accomplishments and Research Findings

- Developed sensors and integrated them into APR cartridges for evaluation.
- Developed new nanomaterials for chemical sensing.

Publications and Abstracts

- King BH, Ruminski AM, Snyder JL, Sailor MJ [2007]. Optical-fibermounted porous silicon photonic crystals for sensing organic vapor breakthrough in activated carbon. Adv Mater *19*:4530–44.

Invited Presentations

- Snyder J [2007]. End-of-service-life sensor systems for personal protective equipment. Scientific conference on chem bio defense research, Timonium, MD, November 13–15.
- Snyder J [2007]. End-of-service life sensors and models. Technical Support Working Group, Personal Protective Equipment Conference, Fort Lauderdale, FL, November 27–29.
- Snyder J [2008]. Chemically modified porous silicon for optical sensing of organic vapor breakthrough in activated carbon filters. Nanomaterials for Defense conference, Arlington, VA, April 21–24.
- Snyder J [2008]. End-of-service life sensor development. International Isocyanate Institute (III) America analytical subcommittee meeting, Boulder, CO, June 9.
- Snyder J [2008]. Jetted nanoparticle chemical sensor circuits for respirator end-of-service-life detection. 12th International meeting on Chemical Sensors, Columbus, OH, July 13–16.
- Snyder J [2008]. A cartridge simulator for testing end-of-service life indicators. International Society for Respiratory Protection, 14th International Conference, Dublin, Ireland, September 14–18.

Project 43. Global Harmonization of Exposure Measurement and Exposure Mitigation Approaches for Nanomaterials

Principal Investigator: Vladimir Murashov, Ph.D.
Project Duration: FY 2006–2008

Critical Topic Areas: international activities in safety and health, including exposure assessment, risk assessment, recommendations and guidance, and communication and information.

Accomplishments and Research Findings

- Coordinated the international and interagency activities of the NIOSH Nanotechnology Research Program.
- Led the development and publication of ISO TR/12885:2008 Health and safety practices in occupational settings relevant to nanotechnologies.
- Provided technical expertise for several ISO TC229 projects.
- Chaired the Organization for Economic Cooperation and Development Working Party for Manufactured Nanomaterials Steering Group 8, exposure measurement and exposure mitigation.
- Co-led the WHO project on best practices for safe handling of nano- materials.
- Fostered collaboration in nanotechnology research with the following:
  - EPA
  - Consumer Products Safety Commission
  - NASA
  - National Institute of Standards and Technology
  - European Comission
  - Australia
- Represented NIOSH on several national and international working groups on nanotechnology safety and health.
- Represented NIOSH on the National Nanotechnology Initiative interagency subcommittee and its working groups.
- Presented overviews of the NIOSH Nanotechnology Research Program to outside stakeholders.

Publications and Abstracts

- Murashov VV, Howard J [2007]. Biosafety, occupational health, and nanotechnology. Appl Biosafety *12*(3): 158–167
- Murashov V, Howard J [2008]. The U.S. must help set internationalstandards for nanotechnology. Nature Nanotechnol *3*:635–636.
- Schulte PA, Geraci CL, Zumwalde R, Hoover M, Castranova V, Kuempel E, Murashov V, Vainio H, Savolainen K [2008]. Sharpening the focus on occupational safety and health of nanotechnology in the workplace. Scand J Work Environ Health *34*(6):471–478.
- ISO [2008]. Health and safety practices in occupational settings relevant to nanotechnologies. Geneva, Switzerland: International Organization for Standardization, ISO/TR– 12885.
- NIOSH [2008]. Nanotechnology: should carbon nanotubes be handled in the workplace like asbestos? Cincinnati, Ohio: U.S. Department of Health and Human Services, Centers for Disease Control, National Institute for Occupational Safety and

Health, NIOSH Science Blog May. [www.cdc.gov/niosh/blog/nsb052008_nano .html].

- Murashov VV [in press]. Occupational exposure to nanomedical applications. Wiley Interdisciplinary Reviews: nanomedicine.

Government Publications

- NSTC [2007]. The national nanotechnology initiative strategic plan. By Nanoscale science, engineering, and technology subcommittee, committee on technology. National Science and Technology Council, Arlington, VA. [http://www.nano. gov/NNI_Strategic_Plan_2007.pdf].
- NSTC [2007]. Prioritization of environmental, health, and safety research needs for engineered nanoscale materials. An interim document for public comment. By Nanoscale science, engineering, and technology subcommittee, committee on technology. National Science and Technology Council, Arlington, VA.
- NSTC [2008] .Strategy for nanotechnology-related environmental health and safety research. Nanoscale science, engineering, and technology subcommittee, committee on technology. National Science and Technology Council, Arlington, VA. [http://www.nano.gov/NNI_ EHS_Research_Strategy.pdf].

Invited Presentations

- Murashov V [2007]. Nanotechnology health and safety concerns— NIOSH Nanotechnology Research Program. Intertech-PIRA workshop, Washington, DC, February 8.
- Murashov V [2007]. NIOSH Nanotechnology Research Program. Navy occupational health and safety preventive medicine conference, Hampton, VA, March 20.
- Murashov V [2007]. Safe handling of nanotechnology. OSHA/NIOSH Journal Club, Washington, DC, March 29.
- Murashov V [2007]. Nanotechnology health and safety concerns. Meeting of National Aeronautics and Space Administration Code 500 safety awareness campaign, Greenbelt, MD, April 18.
- Murashov V [2007]. NIOSH Nanotechnology Research Program. Conference on Nanotech 2007, Santa Clara, CA, May 21.
- Murashov V [2007]. Nanotechnology: occupational health and safety. Conference on risk assessment for nanomaterials, Cambridge, MA, May 29.
- Murashov V [2007]. Social studies at NIOSH. The Joint Wharton-CHF symposium on social studies of nanotechnology, Philadelphia, PA, June 8.
- Murashov V [2007]. NIOSH Nanotechnology Research Program. Annual environmental health and exposure assessment workshop, Washington, DC, August 17.
- Murashov V [2007]. NIOSH Nanotechnology Research Program. Interagency workshop on the environmental implications of nanotechnology, Washington, DC, September 5.

- Murashov V [2007]. Considerations for selecting standard materials for occupational safety and health. Interagency workshop on standards for environmental health and safety research needs for engineered nanoscale materials, Gaithersburg, MD, September 12.
- Murashov V [2007]. Nanotechnology health and safety concerns. American Society of Safety Engineers, Region VI PDC, Myrtle Beach, SC, September 20.
- Murashov V [2007]. Nanotechnology: occupational health and safety. Mid-Atlantic Regional conference on occupational medicine, Baltimore, MD, October 13.
- Murashov V [2007]. Nanotechnology: occupational safety and health. NanoApplications summit—CleanTech Day, Cleveland, OH, October 24.
- Murashov V [2007]. Nanotechnology: occupational safety and health. State-of-the-Art Conference/International Conference on Health Care Worker Health, Vancouver, British Columbia, Canada, October 26.
- Murashov V [2008]. Nanotechnology and risk. California State University in Northridge, Environmental and Occupational Health Nanotech symposium, Northridge, CA, February 20.
- Murashov V [2008]. International Organization for Standardization TC146/SC2. International Organization for Standardization workshop on documentary standards for measurement and characterization in nanotechnologies, Gaithersburg, MD, February 26.
- Murashov V [2008]. OECD WPMN SG8 Cooperation on exposure measurement and exposure mitigation. International Organization for Standardization workshop on documentary standards for measurement and characterization in nanotechnologies, Gaithersburg, MD, February 26.
- Murashov V [2008]. Exposure assessment and control practices for nanoparticles. University of California, San Diego, forum, San Diego, CA, March 6.
- Murashov V [2008]. Nanotechnology and risk. Joint BIONET/BSAFsymposium on current topics in biosafety, La Jolla, CA, March 7.
- Murashov V [2008]. Nanotechnology and risk. 2008 Controlled Envi-ronment Testing Association conference, San Antonio, TX, April 14.
- Murashov V [2008]. Nanotechnology: exposure assessment. International symposium on risk assessment of manufactured nanomaterials, Tokyo, Japan, April 23.
- Murashov V [2008]. Nano health, safety and environment and workerprotection. Second Jisso international forum, Atlanta, GA, May 22.
- Murashov V [2008]. Nanotechnology: occupational safety and health. Workshop on biological interaction of engineered nanomaterials: environmental, safety, and health issues of military concern, WPAFB, Dayton, OH, June 25.
- Murashov V [2008]. NIOSH Nanotechnology Research Program. Nanoscale Science, Engineering and Technology subcommittee meeting, Arlington, VA, July 15.
- Murashov V [2008]. NIOSH Nanotechnology Research Program. Conference on sustainable nanotechnology innovations, products, and toxicology, Fayetteville, AR, August 14.

- Murashov V [2008]. NIOSH Nanotechnology Research Program. State-of-the-Art Conference/International Conference on Health Care Worker Health, Vancouver, British Columbia, Canada, October 27.
- Murashov V [2008]. NIOSH Nanotechnology Research Program. Meeting of Chinese National Institute of Occupational Health and Poison Control, Beijing, China, November 13.
- Murashov V [2008]. NIOSH Nanotechnology Research Program. Meeting of National Center for Nano science and Technology of China, Beijing, China, November 14.
- Murashov V [2008]. Nanotechnology and risk. Metro Washington College of Occupational and Environmental Medicine Meeting, Bethesda, MD, December 3.

APPENDIX A: PROJECTS BY CRITICAL TOPIC AREA

A. Toxicology and Internal Dose

Project 1: Nanotechnology Research Center Coordination
Project 2: Nanotechnology Safety and Health Research Coordination
Project 3: Systemic Microvascular Dysfunction: Effects of Ultrafine vs. Fine Particles
Project 4: Particle Surface Area as a Dose Metric
Project 5: Role of Carbon Nanotubes in Cardiovascular and COPD-Related Diseases
Project 6: Investigations of Multi-Walled Carbon Nanotube Toxicity
Project 7: Pulmonary Toxicity of Metal Oxide Nanospheres and Nanowires
Project 8: WC-Co Nanoparticles in Initiating Angiogenesis by Reactive Oxygen Species
Project 9: Evaluation of the Pulmonary Deposition and Translocation of Nanomaterials
Project 10: Occupational Exposures and Potential Neurological Risks
Project 11: Lung Effects of Resistance Spot Welding Using Adhesives
Project 12: Neurotoxicity after Pulmonary Exposure to Welding Fumes
Project 13: Potential Aneuploidy Following Exposure to Carbon Nanotubes
Project 14: Pulmonary Toxicity of Carbon Nanotube Particles
Project 15: Dermal Toxicity of Nanotube Particles
Project 16: Specific Biomarkers for Unusual Toxicity of Nanomaterials
Project 17: Pulmonary Toxicity of Diesel Exhaust Particles
Project 18: Induction of Lung Fibrosis by Cerium Oxide in Diesel Exhaust
Project 19: Potential Effects of Silicon-based Nanowires on Lung Toxicity
Project 20: Cell-based Assessment for Iron Nanoparticle-Induced Health Risks
Project 21: Assessment of Carbonaceous Materials on Magenicity

B. Measurement Methods

Project 1: Nanotechnology Research Center Coordination
Project 22: Generation and Characterization of Nanoparticles
Project 23: Dustiness of Nanomaterials

Project 24: Nanoaerosol Monitoring Methods

Project 25: Measurement of Nanoscale Carbonaceous Aerosols

Project 26: Nanoparticle Reference Materials for Health Protection

Project 27: Calm Air Chamber and Wind Tunnel Evaluation of Personal Aerosol Samplers for Nanoparticle Exposure Assessment

Project 28: Ultrafine TiO_2 Surface and Mass Concentration Sampling Method

Project 29: Standard Determination of Nanoparticle Size

Project 30: Development and Evaluation of Nanoaerosol Surface Area Measurement Methods

Project 31: Workplace Monitoring of Carbon Nanofibers/Nanotubes

Project 32: Ultrafine Aerosols from Diesel-Powered Equipment

Project 33: Titanium Dioxide and Other Metal Oxides Exposure Assessment Study

C. Exposure Assessment

Project 1: Nanotechnology Research Center Coordination

Project 23: Dustiness of Nanomaterials

Project 24: Nanoaerosol Monitoring Methods

Project 25: Measurement of Nanoscale Carbonaceous Aerosols

Project 31: Workplace Monitoring of Carbon Nanofibers/Nanotubes

Project 33: Titanium Dioxide and Other Metal Oxides Exposure Assessment Study

Project 34: Field Research Team

Project 41: Nanoparticles in the Workplace

Project 43: Global Harmonization of Exposure Measurement and Exposure Mitigation Approaches for Nanomaterials

E. Epidemiology and Surveillance

Project 1: Nanotechnology Research Center Coordination

Project 35: Assessing the Feasibility of Industrywide Exposure and Epidemiology Studies of Workers Exposed to Engineered Nanomaterials

F. Risk Assessment

Project 1: Nanotechnology Research Center Coordination

Project 36: Nanoparticles—Dosimetry and Risk Assessment

Project 43: Global Harmonization of Exposure Measurement and Exposure Mitigation Approaches for Nanomaterials

G. Engineering Controls and Personal Protective Equipment

Project 1: Nanotechnology Research Center Coordination
Project 37: Penetration of Nanoparticles through Respirators
Project 38: Nanoparticle Penetration through Protective Clothing
Project 39: Development of Personal Protective Equipment Ensemble Test Methods

H. Fire and Explosion

Project 1: Nanotechnology Research Center Coordination
Project 23: Dustiness of Nanomaterials

I. Recommendations and Guidance

Project 1: Nanotechnology Research Center Coordination
Project 43: Global Harmonization of Exposure Measurement and Exposure Mitigation Approaches for Nanomaterials

J. Communication and Information

Project 1: Nanotechnology Research Center Coordination
Project 40: Web-based Nanoparticle Information Library Implementation
Project 41: Nanoparticles in the Workplace
Project 43: Global Harmonization of Exposure Measurement and Exposure Mitigation Approaches for Nanomaterials

K. Applications

Project 1: Nanotechnology Research Center Coordination
Project 42: New Sensor Technology Development for ESLI

Appendix B: NIOSH Extramural Nanotechnology Research Activities

Background

NIOSH sponsors research and training outside the Institute through the Office of Extramural Programs (OEP). The creativity and special resources available in the extramural community make these programs an important component in achieving a national goal to have safe jobs and healthy workers. The office also serves to help NIOSH address priorities from the National Occupational Research Agenda (NORA), the NIOSH Research to Practice (r2p) initiative, congressional mandates, and the NIOSH Program Portfolio (http://www.cdc.gov/niosh/programs).

A variety of occupational safety and health-related research grants and cooperative agreements, training grants, and conference grants outside the Institute are managed by OEP. The competitive process for soliciting applications, conducting peer reviews, and for making awards is managed by OEP.

The extramural office works closely with public and private academic institutions, State and Federal agencies, and with small private businesses. Extramural program scientists also consult with intramural scientists. Extramural applicants are encouraged to collaborate with NIOSH scientists when possible; however, extramural projects are managed separately from intramural programs to ensure fair competition, maintain research integrity, and to eliminate financial and intellectual conflicts of interest.

Purpose

Extramural funding of nanotechnology-related research has been undertaken to help increase the knowledge of nanotechnology and manufactured nanomaterials as they relate to occupational safety and health. Research areas supported by NIOSH/OEP include assessment methods for nanoparticles in the workplace, toxicology of manufactured nano- materials, and the use of nanotechnology for improved workplace monitoring. Extramural nanotechnology research adds to the overall development of new information and complements efforts undertaken within the Institute.

Overall NIOSH Extramural Nanotechnology Research Program

From 2001–2008, NIOSH/OEP has committed $5.2 million (M) to extramural nanotechnology research. Ongoing projects currently funded through 2011 account for another $1.4M (Figure 1), which brings the total to about $6.6M. This has been accomplished by the use of non-earmarked research funds, NIOSH program announcements, joint requests for applications with other Federal agencies, and Small Business Innovation Research (SBIR) funds. The office has not received additional funds for nanotechnology research.

Funding Partnerships

OEP has partnered in two joint Requests for Applications (RFAs) for Nanotechnology Research Grants Investigating Environmental and Human Health Issues. The U.S. Environmental Protection Agency's National Center for Environmental Research (NCER) and the National Science Foundation (NSF) participated in FY 2005. The National Institute of Environmental Health Sciences (NIEHS) joined in FY 2006. Funding was available to support Research (R01) grants for 3 years and Exploratory (R21) grants for 2 years.

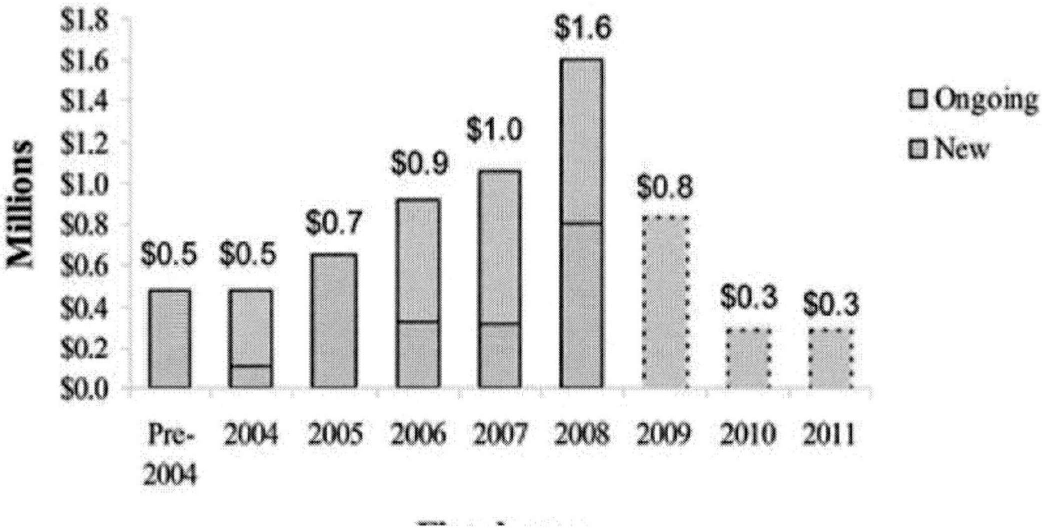

Figure 1. New and ongoing NIOSH/OEP funding in nanotechnology (Pre-2004–2008)

Funding Mechanisms

Figure 2 summarizes the five types of funding mechanisms used by NIOSH for extramural nanotechnology research. SBIR (R43/44) and major research projects (R01) represent the major categories.

Extramural Research Projects and NIOSH Nanotechnology Goals

The major NIOSH goals for nanotechnology research are as follows:

1. Determine if nanoparticles and nanomaterials pose risks for work-related injuries and illnesses.
2. Conduct research to prevent work-related injuries and illnesses by applying nanotechnology products.
3. Promote healthy workplaces through interventions, recommendations, and capacity building.
4. Enhance global workplace safety and health through national and international collaborations on nanotechnology research and guidance.

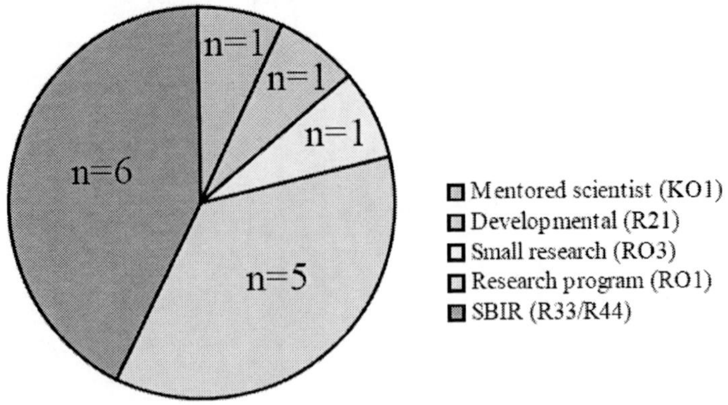

Figure 2. NIOSH/OEP funding (2001–2008) in nanotechnology by Grant Mechanism

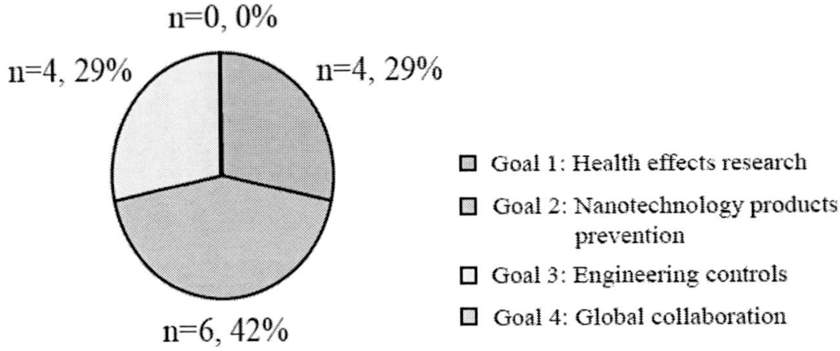

Figure 3. NIOSH/OEP nanotechnology funding by Strategic Goals

Extramural NIOSH funding for nanotechnology research has primarily been related to goals 1–3 (Figure 3). NIOSH/OEP continues to pursue partnerships in sponsoring funding opportunities that involve global collaborations between scientists in the United States and other countries.

Additional Information about Extramural Portfolio

From FY 2001 to FY 2004, OEP funded three R43/44 projects for a total funding commitment of about $950K (Table 1).

In FY 2005, NIOSH funded two nanotechnology research grants through the R01 Program Announcement (Table 1). OEP also began participating in RFAs for Nanotechnology Research Grants Investigating Environmental and Human Health Issues. For the first RFA, 83 applications were received and 19 were recommended for funding. Fourteen of these met NIOSH criteria for relevance to occupational safety and health. Five were in the competitive range for funding consideration and one was funded by NIOSH. EPA funded 14 projects and NSF funded two projects under this RFA.

Table 1. Extramural Nanotechnology Research Funded by NIOSH, 2001–2008

Grant Number	Investigator	Institution	Project title	Start	End	1st year funding	Total funding
Prior to FY-04 NIOSH SBIR							
1R43OH007471-01	Hooker	Nanomaterials Research LLC, Longmont, CO	Novel Hydrogen Sulfide Sensors for Portable Monitors	9/30/2001	3/31/2002	$100,000	$100,000
2R44OH007471-02	Hooker	Synkera Technologies Inc., Longmont, CO	Novel Hydrogen Sulfide Sensors for Portable Monitors	9/16/2003	9/15/2006	$373,400	$750,000
Pre-FY-04 Total						$473,400	$850,000
FY-04 NIOSH SBIR							
1R43OH007963-01A1	Rajagopalan	Nanoscale Materials, Inc., Manhattan, KS	From Nanoparticles to Novel Protective Garments	9/1/2004	5/15/2005	$100, 000	$100,000
FY-04 Total						$100,000	$100,000
FY-05 NIOSH R01 Program Announcement							
1R01OH008282-01A	Kagan	University of Pittsburgh	Lung Oxidative Stress/Inflammation by Carbon Nanotubes	7/1/2005	6/30/2009	$364,000	$1,451,500
FY-05 Co-sponsored RFA (EPA STAR-2005-B1)							
1R01OH008806-01	O'Shaughnessy	University of Iowa	Assessment Methods for Nanoparticles in the Workplace	7/1/2005	6/30/2008	$133,000	$391,200
1R01OH008807-01	Xiong	New York University School of Medicine	Monitoring and Characterizing Airborne Carbon Nanotube Particles	8/1/2005	7/31/2008	$158,000	$395,200
FY-05 Total						$655,000	$2,237,900
FY-06 NIOSH SBIR							
1R43OH008739-01	Thompson	Eltron Research Boulder, CO	Antistatic Paint with Silent Discharge	8/30/2006	8/30/2007	$100,000	$100,000
1R43OH008939-01	Deininger	Synkera Technologies Inc., Longmont, CO	New Nanostructured Sensor Arrays for Hydride Detection	8/1/2006	2/28/2007	$100,000	$100,000
FY-06 Co-sponsored RFA (EPA G2006-STAR F1 to F7)							
1R01OH009141-01	Dutta	Ohio State University	Role of Surface Chemistry in the Toxicological Properties of Manufactured Nanoparticles	9/1/2006	8/31/2009	$120,000	$360,200
FY-06 Total						$320,000	$560,200
FY-07 NIOSH Exploratory/Developmental Research (R21) Program Announcement							
1R21OH009114-01	Volckens	Colorado State University	A High-Flow Personal Sampler for Inhalable Aerosol	8/1/2007	7/31/2009	$207,200	$421,000
FY-07 Career Development Grants (K01) Program Announcement							
1K01OH009255-01	Peters	University of Iowa	Personal Exposure to Engineered Nanoparticles	9/1/2007	8/31/2010	$106,900	$321,000
FY-07 Total						$314,100	$742,000
FY-08 Co-sponsored RFA (NIEHS RFA-ES-06-008)							
1R01OH009448-01	Grassian	University of Iowa	An Integrated Approach Toward Understanding the Toxicity of Inhaled Nanomaterials	4/1/2008	3/31/2012	$337,600	$1,222,600
FY-08 NIOSH Small Research Grant (PAR-06-551) Program Announcement							

Table 1 (Continued)

1R03OH009381	Volckens	Colorado State University	A Personal Sampler for Assessing Inhaled Nanoparticle Exposures	7/1/2008	6/30/2010	$71,500	$143,000
FY-08 NIOSH SBIR							
2R44OH007963-02	Rajagopalan	Nanoscale Materials, Inc., Manhattan, KS	From Nanoparticles to Novel Protective Garments	7/1/2008	6/30/2010	$387,700	$750,000
FY-08 Total						$796,800	$2,115,600
Grand Total							$6.6M

In FY 2006, 81 applications were received in response to the joint RFA on environmental and human health issues. Six of these met NIOSH criteria for relevance and three were in the competitive range for funding consideration. NIOSH was able to fund one application (Table 1). EPA funded 21 projects, NSF funded four, and NIEHS funded three projects under this RFA. NIOSH also funded two SBIR grants on nanotechnology in FY 2006.

In FY 2007, NIOSH funded a 3-year career development (mentored scientist) grant involving research on personal exposure to nanoparticles at the University of Iowa. In addition, a 2-year exploratory/developmental research project was funded at Colorado State University. NIOSH also participated in RFA-ES-06-008 Manufactured Nanomaterials: Physico-chemical Principles of Biocompatibility and Toxicity, which was jointly sponsored by NIEHS, EPA, and NIOSH.

In FY 2008, NIOSH funded a major research grant (R01) from the joint RFA on Manufactured Nanomaterials (Physico-chemical Principles of Biocompatibility and Toxicity). This project is being conducted at the University of Iowa. NIOSH also funded a small research grant (R03) at Colorado State University and an SBIR grant at Nanoscale Materials, Inc. in Kansas.

In FY 2009, nanotechnology related research proposals submitted to standing program announcements will be considered for funding. NIOSH plans to continue collaborative efforts with EPA/NCER, NSF, NIH/NIEHS, and other international agencies to support nanotechnology research with occupational safety and health implications. OEP will continue to consult with the NIOSH Nanotechnology Research Center regarding needs and future directions for nanotechnology research.

Extramural Project Descriptions

Summaries of the extramural projects funded by NIOSH/OEP are included as part of this portfolio update. Contact information for the principal investigators of the projects is provided in Table 2 to encourage collaborative scientific efforts among researchers.

Current NIOSH Nanotechnology Related Information

Extramural investigators interested in pursuing nanotechnology studies related to occupational health and safety can learn more about the interests of NIOSH in this area by visiting the following Web pages: http://www.cdc.gov/niosh/r2p/.

NIOSH has identified 10 critical topic areas to guide in addressing knowledge gaps, developing strategies, and providing recommendations. The key role for NIOSH in conducting and partnering in research on occupational exposures to nanomaterials is noted in a new strategic plan under the National Nanotechnology Initiative.

Information about the NIOSH goal to transfer research findings, technologies, and information into prevention practices and products in the workplace is provided at the Web site This emphasis on Research to Practice (r2p) is intended to reduce occupationally-related illness and injury by increasing the use of findings from NIOSH-funded research in the workplace.

NIOSH conducts a wide range of efforts in the areas of research, guidance, information transfer, and public service. Additional information about the diverse activities of NIOSH can be found on the NIOSH home page (http://www.cdc.gov/niosh/homepage.html) and the program portfolio Web site (http://www.cdc.gov/niosh/programs/).

The Nanoscale Science, Engineering, and Technology (NSET) Subcommittee of the National Science and Technology Council's Committee on Technology, with support from the National Nanotechnology Coordination Office, has released a new NNI Strategic Plan as called for in the 21st Century Nanotechnology Research and Development Act (Public Law 108-153) of 2003. This plan updates and replaces the NNI Strategic Plan of December 2004 and is available at http://www.nano.gov/NNI_Strategic_Plan_2007.pdf . NIOSH participates as a member organization on this subcommittee.

Extramural Nanotechnology Research Project Summaries

The project summaries included here contain the publicly available information from the NIH CRISP database (http://crisp.cit.nih.gov/). For additional information, please contact the Office of Extramural Programs. Investigator contact information is provided in Table 2 to encourage collaborative scientific efforts between researchers. Contact information for Dr. Hooker was not available. Project summaries are listed in the order they appear in Table 1.

HOOKER 7471 (R43, R44)

Novel Hydrogen Sulfide Sensors for Portable Monitors

Background

The primary objective for this project is the design, development, and demonstration of better sensor technology for the detection of hydrogen sulfide. Hydrogen sulfide is a highly toxic, colorless, flammable gas that reacts with enzymes that inhibit cell respiration.

Table 2. Nanotechnology Principal Investigators Funded by NIOSH/ OEP, 2001–2008

Prabir K. Dutta The Ohio State University Department of Chemistry 1960 Kenny Road Columbus, OH 43210 (614) 292–4532 (t)/(614) 292–1685 (f) dutta.1@osu.edu 1R01OH009141 (2006)	**Thomas M. Peters** University of Iowa Department of Occupational and Environmental Health 100 Oakdale Campus, 102 IREH Iowa City, IA 52242–5000 (319) 335–4436 (t)/ (319)335–4225 (f) thomas-m-peters@uiowa.edu
Debra J. Deininger Synkera Technologies, Inc. 2021 Miller Drive, Suite B Longmont, CO 80501 (720) 494–8401 (t)/(720) 494–8402 (f) ddeininger@synkera.com 1R43OH008939-01 (2006)	**Shymala Rajagopalan** NanoScale Materials, Inc. 1310 Research Park Dr. Manhattan, KS 66502 (785) 537–0179 (t)/(785) 537–0226 (f) SRAJAGOPALAN@NANOACTIVE.COM www.nanoactive.com 1R43OH007963-01A1 (2004)
Vicki H. Grassian Professor and Institute Director Nanoscience & Nanotechnology Institute University of Iowa The University of Iowa Iowa City, IA 52242 (319) 384–3292 (t)/(319) 353–1115 (f) vicki-grassian@uiowa.edu 1R01OH009448-01 (2008)	**Patrick Thomas O'Shaughnessy** University of Iowa Department of Occupational and Environmental Health 100 Oakdale Campus, 137 IREH Iowa City, IA 52242–5000 (319) 335–4202 (t)/(319) 335–4225 (f) Patrick-Oshaughnessy@uiowa.edu 1R01OH008806-01 (2005)
Professor Valerin E. Kagan, Ph.D., D.Sc. University of Pittsburgh Department of Environmental and Occupational Health Bridgeside Point 100 Technology Drive, Suite 350 Pittsburgh, PA 15129 (412) 624–9479 (t)/(412) 624–9361 (f) kagan@pitt.edu 1R01OH008282-01A (2005)	**Joel S. Thompson** Eltron Research 4600 Nautilus Court South Boulder, CO 80301–3241 (303) 530–0263 Ext. 118 (t); (303) 530–0264 (f) eltron@eltronresearch.com 1R43OH008739-01 (2006)
Dr. John Volckens 153 Environmental Health Building Colorado State University Fort Collins, CO 80523–1681 (970) 491–6341 (t); (970) 491–2940 (f) John.Volckens@colostate.edu 1R03OH009381	**Judy Xiong** New York University School of Medicine Environmental Medicine 57 Old Forge Road Tuxedo, NY 10987 (845) 731–3627 (t)/(845) 351–5472 (f) xiongj@env.med.nyu.edu 1R01OH008807-01 (2005)
Stephen Williams Synkera Technologies, Inc. 2021 Miller Drive, Suite B Longmont, CO 80501 (720) 494–8401 (t)/(720) 494–8402 (f) swilliams@synkera.com 2R44OH007471-02 (2003)	

At high concentrations, hydrogen sulfide can shut off the lungs, while lower concentrations can burn the respiratory tract and cause eye irritation.

This gas is encountered in a wide range of industries, and a number of standards have been established for occupational exposure. The OSHA permissible exposure limit (PEL) is 10 ppm, the short term exposure limit (STEL) is 15 ppm, and exposures of 300 ppm or greater are considered immediately dangerous to life and health (IDLH). Because of the potential for adverse health effects at low concentrations, the industrial hygiene community is continually seeking improved performance from hydrogen sulfide sensors. Specific requirements include reliable and accurate detection in real time, quantitative measurement capabilities, low purchase and life cycle costs, and low power consumption (for portability). Sensors meeting these requirements will find numerous applications within the health and safety field. In addition, several potential spin-off opportunities are available in leak detection, emission monitoring, and process control. We will use alternative ceramic oxide materials and a unique multi-layer fabrication process to accomplish the objectives of this project. The work plan includes optimization of the sensor materials, sensor element fabrication, sensor element packaging, in-house and external evaluation of the sensors, and establishing the foundation for new instrument development. The ultimate aim is a low-cost, low-power sensor that can be used in a new type of personal monitor. The envisioned monitor is a low profile, credit card sized "smart-card" that not only alerts the wearer when unsafe concentrations have been encountered but also tracks cumulative individual worker exposure to a particular toxic gas species.

Progress

The early portions of this project focused on developing nano-structured sensor materials and morphologies targeted toward chemiresistive-based sensors. This work resulted in a commercial sensor that is being incorporated into the fixed-system H2S detection products of multiple partnering instrument manufacturers. The middle portion of the project focused on developing solid-state electrochemical sensors using micro and nano-sized morphologies and structures. This has resulted in a working sensor currently being incorporated into an inexpensive, portable personal monitor as described above.

RAJAGOPALAN (7963-R43)

From Nanoparticles to Novel Protective Garments

Background

The overall objectives of this collaborative Phase I research between NanoScale and Gentex Corp. are to (1) investigate the use of highly adsorbent and reactive nanoparticles in protective garments and (2) create and test new materials for use in the production of protective clothing. During routine chemical use, it is not always apparent when exposure occurs. Many chemicals pose invisible hazards and offer no warnings. More importantly, terrorists and saboteurs use a variety of toxic industrial chemicals to create improvised explosives, chemical agents, and poisons. When hazardous materials are released, protective clothing is critical in guarding against the effects of toxic or corrosive products that could enter the body by inhalation or skin absorption and cause adverse effects.

This project seeks to (a) establish the feasibility of incorporating highly adsorbent and reactive nanoparticles into lightweight, permeable textiles and (b) evaluate the utility of the resultant fabric as protective clothing using standard industry testing procedures. These novel protective garments will be tailored toward personnel associated with Federal, State, or local emergency agencies as well as fire fighters and civilian first responders.

To achieve the overall objective, reactivity of selected nanoparticle formulations to various toxic industrial chemicals will be explored by use of a quartz spring balance to determine sorptive capacity. Based on the outcome of this research, a single reactive nanoparticle formulation will be chosen for use in fabrics. The selected nanoparticle formulation will then be incorporated into suitable fabrics using two established techniques. Next, fabric test swatches will be evaluated for a number of criteria using industry recognized ASTM test methods. Finally, the top four nanoparticle embedded fabrics will be tested for physical and chemical resistance against two representative toxic chemicals using a standard ASTM procedure.

KAGAN (8282)

Lung Oxidative Stress/Inflammation by Carbon Nanotubes

Background

Specific Aim 1 is to establish the extent to which SWCNT alone are pro-inflammatory to lung cells and tissue and characterize the role of iron in these effects using genetically manipulated cells and animals as well as antioxidant interventions.

Specific Aim 2 is to determine the potential for SWCNT and microbial stimuli to synergistically interact to promote macrophage activation, oxidative stress, and lung inflammation.

Specific Aim 3 is to reveal the extent to which SWCNT are effective in inducing apoptosis and whether apoptotic cells exert their macrophage-dependent anti-inflammatory potential during in vitro and in vivo SWCNT exposure. The project involves a team of inter-disciplinary scientists with expertise in redox chemistry/biochemistry, cell and molecular biology of inflammation and its interactions with microbial agents, and pulmonary toxicology of nanoparticles.

Progress

Two types of single walled carbon nanotubes were used (iron-rich and iron-stripped) to study their interactions with RAW 264.7 macrophages. Following ultrasonication of the SWCNT to separate strands, neither types were able to generate intracellular production of superoxide radicals or nitric oxide macrophages observed by flow-cytometry and fluorescence microscopy. SWCNT with different iron content displayed different redox activity in a cell-free model system. In the presence of microbial (zymosan)-stimulated macrophages, non-purified, iron-rich SWCNT were more effective in generating hydroxyl radicals than purified SWCNT. The presence of iron in SWCNT may be important in determining redox-dependent responses of macrophages. Dose and time-dependence studies of inflammatory responses in mice using pharyngeal aspiration of SWCNT demonstrated that SWCNT elicited unusual pulmonary effects in C57BL/6 mice that combined a robust but acute inflammation with early onset yet progressive fibrosis and granulomas. It was demonstrated that occupationally relevant dose-dependent affects of SWCNT may exert toxic effects in the lung of exposed animals in vivo. SWCNT induced inflammation, and exposure caused altered pulmonary function and microbial stimulation. Clearance from the lungs of SWCNT exposed mice was compromised. An unusual and robust inflammatory and fibrogenic response was correlated with the progression of oxidative stress and apoptotic signaling. Not only are toxic effects of SWCNT important to consider but also the role of transitional metals, particularly iron should be investigated.

Recent Publications

Shvedova AA, Fabisiak JP, Kisin ER, Murray AR, Roberts JR, Tyurina YY, Antonini JM, Feng WH, Kommineni C, Reynolds J, Barchowsky A, Castranova V, Kagan VE [2007].Sequential Exposure to carbon nanotubes and bacteria enhances pulmonary inflammation and infectivity. Am J Respir Cell Mol Biol.

Abstract

Realistic exposures to carbon nanotubes (CNT) may occur in conjunction with other pathogenic impacts (microbial infections) and trigger enhanced responses. We evaluated interactions between pharyngeal aspiration of SWCNT and bacterial pulmonary infection of C57BL/6 mice with Listeria monocytogenes (LM). Mice were given SWCNT (0, 10, and 40 microg/mouse) and exposed to LM (10(3) bacteria/mouse) 3 days later. Sequential exposure to SWCNT/LM amplified lung inflammation and collagen formation. Despite this robust inflammatory response, SWCNT pre-exposure significantly decreased the pulmonary clearance of LM measured 3–7 days after microbial infection vs PBS/LM treated mice. Decreased bacterial clearance in SWCNT pre-exposed mice was associated with decreased phagocytosis of bacteria by macrophages and a decrease in nitric oxide production by these phagocytes. Pre-incubation of naive alveolar macrophages with SWCNT in vitro also resulted in decreased nitric oxide generation and suppressed phagocytizing activity towards LM. Failure of SWCNT-exposed mice to clear LM led to a continued elevation in nearly all major chemokines and acute phase cytokines into the later course of infection. In SWCNT/LM exposed mice, BAL PMNs, AMs, and lymphocytes as well as LDH level were increased compared to mice exposed to SWCNT or LM alone. In conclusion, enhanced acute inflammation and pulmonary injury with delayed bacterial clearance after SWCNT exposure may lead to increased susceptibility to lung infection in exposed populations.

Shvedova AA, Kisin ER, Murray AR, Gorelik O, Arepalli S, Castranova V, Young SH, Gao F, Tyurina YY, Oury TD, Kagan VE [2007].Vitamin E deficiency enhances pulmonary inflammatory response and oxidative stress induced by single-walled carbon nanotubes in C57BL/6 mice. Toxicol Appl Pharmacol *221* (3):339–348

Abstract

Exposure of mice to single-walled carbon nanotubes (SWCNTs) induces an unusually robust pulmonary inflammatory response with an early onset of fibrosis, which is accompanied by oxidative stress and antioxidant depletion. The role of specific components of the antioxidant protective system (specifically vitamin E, the major lipid-soluble antioxidant) in the SWCNT-induced reactions has not been characterized. We used C57BL/6 mice, maintained on vitamin E-sufficient or vitamin E-deficient diets, to explore and compare the pulmonary inflammatory reactions with aspired SWCNTs. The vitamin E-deficient diet caused a 90-fold depletion of alpha-tocopherol in the lung tissue and resulted in a significant decline of other antioxidants (GSH, ascorbate) as well as accumulation of lipid peroxidation products. A greater decrease of pulmonary antioxidants was detected in SWCNT-treated vitamin E-deficient mice compared with controls. Lowered levels of antioxidants in vitamin E-deficient mice were associated with a higher sensitivity to SWCNT-induced acute inflammation (total number of inflammatory cells, number of polymorphonuclear leukocytes, released LDH, total protein content and levels of pro- inflammatory cytokines, TNF-alpha and IL-6) and enhanced profibrotic responses (elevation of TGF-beta and collagen deposition). Exposure to SWCNTs markedly shifted the ratio of cleaved to full-length extracellular superoxide dismutase (EC-SOD). Given that pulmonary levels of vitamin E can be manipulated through diet, its effects on SWCNTinduced inflammation may be of practical importance in optimizing protective strategies.

O'SHAUGHNESSY (8806)

Assessment Methods for Nanoparticles in the Workplace

Background

The primary objectives are to provide the scientific community and practicing industrial hygienists with verified instruments and methods for accurately accessing airborne concentrations of nanoparticles, and to assess the efficacy of respirator use for controlling nanoparticle exposures.

We will satisfy these objectives through a combination of laboratory and field-based studies centered on the following specific aims: (1) identify and evaluate methods to measure airborne nanoparticle concentrations, (2) characterize nanoparticles using a complementary suite of techniques to assess their surface and bulk physical and chemical properties, and (3) determine the collection efficiency of commonly-used respirator filters when challenged with nanoparticles.

Progress

Several methods were used to aerosolize nanoparticles from bulk powders in the laboratory. An apparatus was developed to inject the aerosol into a mainflow of dry, filtered air through a charged neutralizing device. The amount and size distribution of the aerosol in the chamber is sampled with a scanning mobility particle sizer; samples from the chamber are also being analyzed by TEM. Although the primary particle size of these powders average 20nm, an aerosol with a median size of 120 is generated. These findings have significance in occupational settings since agglomeration of the particles in this size range will have consequences in pulmonary deposition and respirator filtration. Nanosized particles were also found as contaminants in the water used. A variety of instruments are being compared for use in the field studies of nanoparticle exposure levels in two facilities, one in MN and one in TX.

Recent Publication

Peters TM, Elzey S, Johnson R, Park H, Grassian VH, Maher T, O'Shaughnessy P [2009]. Airborne monitoring to distinguish engineered nanomaterials from incidental particles for environmental health and safety. J Occup Environ Hyg *6*:2,73–8 1.

Abstract

Two methods were used to distinguish airborne engineered nanomaterials from other airborne particles in a facility that produces nano-structured lithium titanate metal oxide powder. The first method involved off-line analysis of filter samples collected with conventional respirable samplers at each of seven locations (six near production processes and one outdoors). Throughout most of the facility and outdoors, respirable mass concentrations were low (<0.050 mg/m^3) and were attributed to particles other than the nanomaterial (<10% by mass titanium determined with inductively coupled plasma atomic emission spectrometry). In contrast, in a single area with extensive material handling, mass concentrations were greatest (0.118 mg/m^3) and contained up to 39% ± 11% lithium titanium, indicating the presence of airborne nanomaterial. Analysis of the filter samples collected in this area by transmission electron microscope and scanning electron microscope revealed that the airborne nanomaterial was associated only with spherical aggregates (clusters of fused 10–80 nm nanoparticles) that were larger than 200 nm. This analysis also showed that nanoparticles in this area were the smallest particles of a larger distribution of sub- micrometer chain agglomerates likely from welding in an adjacent area of the facility. The second method used two hand-held, direct-reading, battery-operated instruments to obtain a time series of very fine particle number (<300 nm), respirable mass, and total mass concentration, which were then related to activities within the area of extensive material handling. This activity-based monitoring showed that very fine particle number concentrations (<300 nm) had no apparent correlation to worker activities, but that sharp peaks in the respirable and total mass concentration coincided with loading a hopper and replacing nanomaterial collection bags. These findings were consistent with those from the filter-based method in that they demonstrate that (1) airborne nanoparticles in this facility are dominated by "incidental" sources (e.g., welding or grinding), and (2) the airborne "engineered" product is predominately composed of particles larger than several hundred nanometers. The methods presented here are applicable to any occupational or environmental setting in which one needs to distinguish incidental sources from engineered product.

XIONG (8807)

Monitoring and Characterizing Airborne Carbon Nanotube Particles

Background

The proposed research will develop a comprehensive yet practical method for sampling, quantification, and characterization of carbon nanotube (CNT) particles in air. This method will be capable of (1) classifying sampled particles into three categories: tubes, ropes (bundles of single-walled CNTs bounded by Van der Waals attraction force), and non-tubular particles (soot, metal catalysts, and dust, etc.) and (2) measuring the number concentrations and size distributions for each type, and the shape characters (diameter, length, aspect ratio and curvature) of CNTs.

The method will use available instrumentation to build an air monitoring system that is capable of sampling and sizing airborne CNT particles in a wide size range by using a 10-stage, micro-orifice uniform deposit impactor (MOUDI) and an integrated diffusion battery previously developed in this laboratory. Successful completion of this project will produce a validated method for sampling airborne.

CNTs in workplaces; and a practical method using atomic force microscopy image analysis technology to classify sampled CNT particles by type, and quantifying and characterizing each type separately. These methods are needed to determine health risks that may result from worker exposure to the various types: CNTs, nanoropes, and non-tubular nanoparticles. The results will also provide a foundation for field and personal sampling devices for CNTs.

Progress

Instrumentation and materials are essentially ready. Years 2 and 3 will focus on method development for sampling, quantifying, and characterizing airborne CNT aerosol particles.

THOMPSON (8739)

Antistatic Paint with Silent Discharge

Background

Proposed is an anti-static paint that can be used to alleviate danger from fires and explosions initiated by static discharge. The innovation for the proposed paint lies in its discharge mechanism, which doesn't require grounding or hydration like other products that are currently on the market. Paint can be sprayed onto surfaces of clothing or equipment, resulting in a lower capacitance or ability to build up charge on the material.

Research Design

A series of formulations will be synthesized and tested on both metallic and plastic substrates.

Relevance to Public Health

Development of the antistatic paint will lead to immediate safety improvements for persons that work with volatile solvents in the fuel, coatings, and plastic industries as well as emergency responders in haz-mat situations. The paint could be useful for protecting the general public from electrostatic initiated explosions at the gas-pump or from uncomfortable static shock. In addition, this paint could lead to substantial savings in the electronics industry, where electrostatic discharge (ESD) events cost millions of dollars each year because of lost products.

DEININGER (8939)

New Nanostructured Sensor Arrays for Hydride Detection

Background

The goal of the proposed project is to develop improved sensors for the detection of hydrides (including arsine, phosphine and diborane) for protecting worker safety and health. Current sensors suffer from severe limitations including lack of selectivity, and limited accuracy and lifetime. An electronic sensor system, capable of automatically warning workers of the presence of one of these toxic gases, would provide a substantial benefit for worker safety and health.

This project will take advantage of advances in nanotechnology, ceramic micromachining, and materials chemistry to create better sensors. These improved sensors will be the basis for improved personal and permanent monitors for increased protection of workers in the semiconductor industry.

Progress

Phase I feasibility studies have shown a significant enhancement in the sensitivity of hydride sensors fabricated from tailored nanostructured materials grown via a templated deposition process inside the pores of anodic alumina. Detection limits of less than 10 ppb were achieved, allowing application of this rugged, low-cost sensor technology to the detection of highly toxic gases (e.g. hydrides) that heretofore were not amenable to reliable detection via this sensor technology.

DUTTA (9141)

Role of Surface Chemistry in the Toxicological Properties of Manufactured Nanoparticles

Background

The objectives of this program are to verify two hypotheses. First, the quantifiable differences in surface reactivity of nanoparticles (as measured by acidity, redox chemistry, metal ion binding and Fenton chemistry) compared with micron-sized particles of similar composition cannot be explained by the increase in surface area alone. Second, the oxidative stress and inflammatory response induced by nanoparticles upon interaction with macrophages and epithelial cells depend on their surface reactivity. The basis of these hypotheses is that nanoparticles contain significantly higher numbers of "broken" bonds on the surface that provide different reactivity as compared to larger particles.

The experimental approach focuses on three classes of manufactured nanoparticles, catalysts (aluminosilicates), titania, and carbon. For the catalysts and titania samples, nanoparticles (< 100 nm) and micron-sized particles of similar bulk composition will be studied. For carbon, carbon black and single walled carbon nanotubes are chosen. Nanoparticles of alumino silicates and titania will be synthesized, whereas the other particles will be obtained from commercial sources. Characterization will involve electron microscopy, surface area, and surface and bulk composition.

Reactivity of well-characterized particles in regards to their acidity, reaction with antioxidants simulating the lung lining fluid, coordination of iron, and Fenton chemistry will be carried out using spectroscopic methods. Particular attention will be paid to surface activation as may exist during manufacturing and processing. In vitro oxidative stress and inflammatory responses upon phagocytosis of the particles by macrophages and pulmonary epithelial cells will form the toxicological/biological end points of the study. Methods include gene array techniques, assays for reactive oxygen species and adhesion molecules on endothelial cells.

VOLCKENS (9114)

A High-Flow Personal Sampler for Inhalable Aerosol

Background

While occupational respiratory diseases inflict a severe toll on our national workforce, existing exposure assessment methods for many respiratory hazards lack sufficient detection sensitivity, preventing the practicing industrial hygienist from determining whether a hazard even exists. The goal of this research is to design a high-flow sampling device engineered to assess inhalable aerosol hazards in the workplace in a physiologically relevant fashion. This device will extend overall method detection limits by considerably increasing sampler flow rate. The design will be inexpensive and readily integrated into the commonly used 37-mm filter cassette. Because of its low cost, accuracy, ease-of-use, and adaptability, this device will have widespread application, especially in developing countries where exposure assessment resources and budgets are limited. Such widespread use was a major design goal for this technology, as increased use of an accurate exposure assessment technology will ultimately result in increased protection of worker health on a national and global scale. This project has three specific aims: (1) To design an inexpensive, modular sampling head that (a) interfaces with a standard 37-mm cassette, (b) operates at a high rate of airflow, 10 Lpm, and (c) meets the criteria for inhalable sampling set by the American Conference of Governmental Industrial Hygienists/International Standards Organization. This high-flow design will proceed from a wide body of research on existing low-flow inhalable devices. (2) Evaluate sampler performance using computational fluid dynamic modeling and to refine the prototype design based on results from these tests. (3) To validate the refined design through laboratory tests in a wind tunnel and field tests in actual occupational environments.

PETERS (9255)

Personal Exposure Monitoring to Engineered Nanoparticles

Background

Worldwide production of engineered nanoparticles is expected to grow from 2,000 metric tons to 50,000 metric tons over the next decade. New industrial processes must be introduced into the workplace to accommodate this growth. Although studies have shown some nanoparticles to be toxic, methods to assess personal exposure do not exist. Knowledge of personal exposure may be particularly important for such small particles because their concentration tends to decay rapidly with distance from a source.

Dr. Peters will conduct laboratory studies to evaluate the precision and accuracy of methods developed by his research group to assess personal exposure to nanoparticles. These methods will then be used to investigate the extent to which workers are exposed to engineered nanoparticles in a facility that produces and handles them. Mixed models will be used to identify the determinants of exposure while controlling for the between- worker (spatial) and within-worker (temporal) variability. Computer-controlled electron microscopy with energy dispersive X-ray detection will be used to further characterize the samples collected in the field study by size, composition, and morphology. These data will be used to apportion exposures to sources. The research proposed in this application is significant, because it will enable direct assessment of personal exposure to nanoparticles on time scales relevant to potential acute and chronic adverse health outcomes. As an outcome of these studies, an understanding of exposures will help to prioritize studies in toxicology, epidemiology, and engineering controls to better protect workers.

The research component will be complemented by a vigorous career development plan that will include the following: (1) formal training in responsible conduct of research, epidemiology, and electron microscopy; (2) regular meetings with the sponsors of this award; (3) participating in group meetings and departmental seminars; (4) presenting results at scientific meetings; and (5) publishing results in peer-reviewed journals. The multidisciplinary team of sponsors will play an active roll in both the research and career development component of this award.

GRASSIAN (9448)

An Integrated Approach toward Understanding the Toxicity of Inhaled Nanomaterials

Background

Manufactured nanomaterials are found in cosmetics, lotions, and coatings, and are used in environmental remediation applications. An opportunity exists for exposure through many different routes, making it necessary to study the health implications of these materials. The primary objective of this research is to fully integrate studies of the physical and chemical properties of commercially manufactured nanoparticles with inhalation toxicological studies of these same nanoparticles to determine properties that most significantly affect nanoparticle toxicity. Our central hypothesis is that nanoparticle physico-chemical properties differ widely among particle types, and certain properties induce adverse health outcomes. Furthermore, we hypothesize that nanoparticle toxicity is influenced by the susceptibility of the individual as well as the presence of other inflammatory substances. We will address these hypotheses through a series of specific study aims designed to establish a relationship between nanoparticle physicochemical properties and health outcomes.

Specific Aim 1: Evaluate nanoparticle chemical composition (bulk and surface) on nanoparticle toxicity in acute and sub-acute exposure studies. Experiments will be designed to investigate nanoparticle composition (bulk and surface) before, during, and after inhalation exposure studies.

Specific Aim 2: Determine the impact of nanoparticle physical morphology (agglomeration size, agglomeration state, and nanoparticle shape) on nanoparticle toxicity. This study will incorporate animal inhalation studies to determine the relationship between nanoparticle agglomerate size and nanoparticle shape on toxicity.

Specific Aim 3: Determine whether (1) pulmonary clearance is impaired by inhaled nanoparticles and (2) impaired clearance increases the risk of pulmonary infection. The pulmonary clearance mechanism, especially the ability of alveolar macrophages to clear microbes or foreign particles, can be impaired by inhaled particulates. We will compare lung clearance rates after inhalation of nanoparticles of different composition.

Specific Aim 4: Compare lung inflammation produced by co-exposure of nanoparticles with other inflammatory substances and relative to the nanoparticles alone. We plan to evaluate synergistic effects with other common aerosols present in the indoor and outdoor environments including endotoxins and sulfate aerosols (e.g. ammonium sulfate).

VOLCKENS (9381)

A Personal Sampler for Assessing Inhaled Nanoparticle Exposures

Background

Elucidation of potential nanoparticle exposure-disease relationships requires the development of an adequate exposure assessment method, as exposure assessment provides key input into subsequent epidemiologic and toxicological investigation. The absence of a personal exposure assessment method for nanoparticle aerosols prevents progress towards understanding potential nanoparticle exposure-disease relationships. The goal of this work is to develop an accurate, sensitive, and specific method to assess personal exposures to engineered nanoparticles. The method will assess the inhalation route of exposure, and hence, the measurement of nanoparticle concentrations in air.

To quantify concentrations of airborne nanoparticles, a measurement must be able to segregate the aerosol by size (to capture only particles smaller than 100 nm) and count these particles following size segregation. Also, because inhaled nanoparticles enter strictly through the oral/nasal route, the method should be oriented to measure breathing zone concentrations within the vicinity of the upper torso. Finally, and perhaps most importantly, the method must be able to distinguish engineered nanoparticles from other incidental nanoparticles (i.e., ultrafine particles) that are ubiquitous in workplace and ambient air.

Successful completion of this research will establish a method to estimate human exposures to engineered nanoparticles in workplace environments. Such a method is critical to the establishment of nanoparticle dose-response relationships, as current methods lack both specificity and sensitivity. Results from this research can also be translated to the larger realm of health-related air pollution outside of the workplace (e.g., indoor air, ambient air pollution), both nationally and internationally.

RAJAGOPALAN (7963-R44)

From Nanoparticles to Novel Protective Garments

Background

The goal of this project is to develop breathable, lightweight, and disposable chemical protective clothing for use by personnel associated with law enforcement agencies, emergency medical services, medical/triage facilities, and other Federal, State, or local emergency agencies as well as by fire fighters and civilian first responders. This goal will be accomplished by producing a lightweight, chemical protective garment material. The material will have a limited number of distinct layers, that can be assembled into a breathable, protective garment. The layers impart at least one key property to the composite material. Non-breathable materials retard the human body's process of heat dissipation normally achieved through the evaporation of perspiration. Without significant transmission of water vapor or breathability, prolonged use of non-breathable materials can result in intolerable discomfort and even death to a person wearing garments made from these materials. The novelty of the proposed approach is the development of granulated reactive NanoActive(r) sorbents containing composite layered textile material. The material protects the sorbent from contact with and contamination by liquids and has good vapor permeability, thus providing for passage of air and water vapor as well as sorption of harmful and noxious vapors and gases. Based on the outcome of Phase I research, it is clear that granulated mixed metal oxide nano-formulations are required for a broad- spectrum reactivity. As a result, this Phase II project will begin by developing methodologies for granulation of mixed metal oxides based on NanoActive materials. This will be followed by performance screening of formulations against toxic industrial chemicals and chemical warfare agent simulants by a breakthrough testing procedure. The down selected formulations will then be incorporated into down selected textile materials with optimum air and water vapor permeable properties. Next, these composite fabric test swatches will be evaluated for a number of criteria using industry-recognized ASTM test methods. The top four granulated nanoparticles embedded fabrics will be tested for chemical resistance against four toxic chemicals using a standard ASTM procedure. The project will conclude by pilot studies involving scale-up of granulated nanoparticles and composite textile preparation followed by production of prototype chemical protective clothing. Relevance major deficiency of the available chemical protective garments is the failure to provide desired levels of toxic vapor sorption and still maintain sufficient transfer of heat and moisture to keep a wearer cool and comfortable. Permeable textiles containing highly reactive NanoActive(r) materials offer promise of protection and comfort. Public health relevance: major deficiency of the available chemical protective garments is the failure to provide desired levels of toxic vapor sorption and still maintain sufficient transfer of heat and moisture to keep a wearer cool and comfortable. Permeable textiles containing highly reactive NanoActive(r) materials offer promise of protection and comfort.

INDEX

A

B

C

D

H

I

J

K

L

M

N

O

P

Q

R

U

V

W

X